# Molecular Modeling and Multiscaling Issues for Electronic Material Applications

Nancy Iwamoto · Matthew M. F. Yuen
Haibo Fan
Editors

# Molecular Modeling and Multiscaling Issues for Electronic Material Applications

 Springer

Nancy Iwamoto
Honeywell Specialty Materials
Moffett Park Drive 1349
Sunnyvale, CA 94089
USA
e-mail: nancy.iwamoto@honeywell.com

Matthew M. F. Yuen
Department of Mechanical Engineering
Hong Kong University of Science and
  Technology
Clear Water Bay
Kowloon
Hong Kong
e-mail: meymf@ust.hk

Haibo Fan
Philips Investment Co. Ltd.
Philips Innovation Campus Shanghai
No. 2 Building, No. 9 Lane 888
  Tian Lin Road
Shanghai
People's Republic of China
e-mail: hb.fan@philips.com

ISBN 978-1-4899-8837-9
DOI 10.1007/978-1-4614-1728-6
Springer New York Dordrecht Heidelberg London

ISBN 978-1-4614-1728-6 (eBook)

*Cover design:* eStudio Calamar S.L.

Printed on acid-free paper

Springer is part of Springer Science+Business Media (www.springer.com)

# Preface

Performance modeling in electronic materials is not a simple application of one scale. It has been recognized by many reliability experts that failure may start at an atomistic level that cannot be easily measured, which propagates upward to scales that can be measured. As many macroscale modeling techniques depend upon the measured property to perform the calculation, what is still largely missing is the connection between the actual molecular structure and how it contributes to the propagation of the failure. Molecular models are significant where properties can be determined only by knowledge of the composition and molecular structure, and measurements are not needed to perform the calculation, although benchmark/calibration measurements are required to ensure proper techniques are used. However, even for molecular modelers, there is a scale issue ranging from quantum scale electronic effects to longer range molecular effects. Often these issues are coupled, but the molecular modeler has an opportunity to separate root issues from the perspective of the actual chemical structure in order to help the experimentalist narrow down the types of materials to develop or test and the types of measurements that must be made. From this aspect, molecular modeling is very flexible in that it can be used for both materials development as well as diagnostics, thereby accelerating the development cycle.

The chapters in this book were compiled from extended papers delivered at the IEEE EuroSimE conference (also known as the International Conference on Thermal, Mechanical and Multiphysics Simulation and Experiments in Microelectronics and Microsystems), at the Molecular Dynamics session between 2007 and 2010, to be used as illustration and learning tools for the application of molecular modeling in the electronics community. EuroSimE is the only IEEE conference devoted to modeling in the electronics community, and was also the first IEEE community to recognize the usefulness of molecular modeling for electronic material performance by devoting a session just for molecular modeling in their conferences (beginning at the April 16–18, 2007 conference held in London, England). We thought it fitting to start a collection of papers from this conference as the community learns how to apply the tools specific to electronic issues. Although molecular modeling has been historically a well-accepted discipline in the pharmaceutical, petroleum and catalyst industries concerned with chemical performance issues, it has not gained wide acceptance in the materials industry. That is why the acceptance of the discipline by this large modeling community is significant.

This book is separated into five sections, each dealing with different scales and performance issues. We have tried to separate the issues based upon the fundamental model size and the performance features being represented and give you examples ranging from the fundamental quantum mechanics calculations all the way to mesoscale examples which is the initial scaling point deviating from explicit atomistic accounting; however the root interactions in mesoscale models can be attributed directly back to the atomistic scale whether by experimental techniques or by explicit parameterization from atomistic models so is included in the book.

As you will see, the EuroSimE community has been active in exploring all the possible scales available to the molecular modeler and the ways in which molecular modeling may be used to help materials understanding. As in any modeling endeavor because the specific modeling method may change depending upon the material and the device application, the most value out of this collection may be gained by attention to the techniques and philosophies used to gain the performance understanding sought by the author. By no means are these the only modeling methods that can be used, but are the ones that were found to be successful for the author and so are instructive for those both starting out in molecular modeling as well as those experienced in the art.

# Acknowledgments

The editors would like to express their appreciation to IEEE and EuroSimeE for their support in the compilation of papers for this book. In particular we would like to thank the organizers of EuroSimE, Guoqi Zhang (of Philips Lighting and Delft University of Technology), Leo Ernst (of Delft University of Technology) and Olivier de Saint Leger (of Astofo, France), for their encouragement. The editors would also like to thank all the paper authors who have worked very hard on their manuscript to make the publishing of this book a reality. We sincerely hope this book can help promote our common goal of making molecular dynamics simulation a recognized tool in materials and electronic packaging research.

# Contents

**Part I**  **Quantum Mechanics and Molecular Methods:**
**Uses for Property Understanding**

**1**  **Atomistic Simulations of Microelectronic Materials:**
**Prediction of Mechanical, Thermal, and Electrical Properties** . . .   3
V. Eyert, A. Mavromaras, D. Rigby, W. Wolf, M. Christensen,
M. Halls, C. Freeman, P. Saxe and E. Wimmer

**2**  **Using Molecular Modeling Trending to Understand**
**Dielectric Susceptibility in Dielectrics for Display Applications** . .   25
Nancy Iwamoto, Ahila Krishnamoorthy and Edward W. Rutter Jr

**3**  **Understanding Cleaner Efficiency for BARC**
**("Bottom Anti-Reflective Coating") After Plasma Etch**
**in Dual Damascene Structures Through the Practical**
**Use of Molecular Modeling Trends** . . . . . . . . . . . . . . . . . . . . . .   39
Nancy Iwamoto, Deborah Yellowaga, Amy Larson, Ben Palmer
and Teri Baldwin-Hendricks

**Part II**  **Large-Scale Atomistic Methods and Scaling Methods**
**to Understand Mechanical Failure in Metals**

**4**  **Roles of Grain Boundaries in the Strength of Metals**
**by Using Atomic Simulations** . . . . . . . . . . . . . . . . . . . . . . . . . . .   55
Tomotsugu Shimokawa

**5    Semi Emprical Low Cycle Fatigue Crack Growth Analysis
      of Nanostructure Chip-To-Package Copper Interconnect
      Using Molecular Simulation** .............................    77
      S. Koh, A. Saxena, W. D. Van Driel, G. Q. (Kouchi) Zhang
      and R. Tummala

**Part III   Molecular Scale Modeling Uses for Carbon
             Nanotube Behavior**

**6    Thermal Conductivity of Carbon Nanotube Under External
      Mechanical Stresses and Moisture by Molecular
      Dynamics Simulation** ...............................    93
      H. Fan, K. Zhang and M. M. F. Yuen

**7    Influence of Structural Parameters of Carbon Nanotubes
      on their Thermal Conductivity: Numerical Assessment** ........   101
      Bartosz Platek, Tomasz Falat and Jan Felba

**Part IV   Molecular Methods to Understand Mechanical
            and Physical Properties**

**8    The Mechanical Properties Modeling of Nano-Scale Materials
      by Molecular Dynamics** .............................   115
      C. Yuan, W. D. van Driel, R. Poelma and G. Q. (Kouchi) Zhang

**9    Molecular Design of Self-Assembled Monolayer (SAM)
      Coupling Agent for Reliable Interfaces by
      Molecular Dynamics Simulation** .......................   133
      C. K. Y. Wong, H. Fan, G. Q. (Kouchi) Zhang and M. M. F. Yuen

**10   Microelectronics Packaging Materials: Correlating Structure
      and Property Using Molecular Dynamics Simulations** .........   149
      Ole Hölck and Bernhard Wunderle

**Part V   Multiscale Methods and Perspectives**

**11   Investigation of Interfacial Delamination
      in Electronic Packages** ............................   189
      H. Fan and M. M. F. Yuen

**12 A Multiscale Approach to Investigate Wettability of Surfaces with Designed Coating** . . . . . . . . . . . . . . . . . . . . . . . . . . . . . . 203
E. K. L. Chan, H. Fan and M. M. F. Yuen

**13 Glass Transition Analysis of Cross-Linked Polymers: Numerical and Mesoscale Approach** . . . . . . . . . . . . . . . . . . . . . . 213
Sebastian J. Tesarski and Artur Wymyslowski

**14 Investigation of Coarse-Grained Mesoscale Molecular Models for Mechanical Properties Simulation, as Parameterized Through Molecular Modeling** . . . . . . . . . . . . . . . . . . . . . . . . . . . . 231
Nancy Iwamoto

**Abbreviations** . . . . . . . . . . . . . . . . . . . . . . . . . . . . . . . . . . . . . . . 251

**Index** . . . . . . . . . . . . . . . . . . . . . . . . . . . . . . . . . . . . . . . . . . . . . . 255

# Part I
# Quantum Mechanics and Molecular Methods: Uses for Property Understanding

## Introduction

This section contains examples specifically from the quantum scale and examples of how such models are used to gain to larger-scale perspectives. Chapter 1 (Eyert et al.) starts off with very nice overview of molecular modeling techniques, before showing examples of the use of quantum mechanics for property prediction. Significantly Eyert introduces all of the different properties that may be gained by molecular modeling and is an excellent summary of the types of information that may be used in each level. Eyert then continues with specific application examples of both ab initio and classic techniques ranging from electronic to thermal and thermomechanical properties. Chapters 2 and 3 (Iwamoto et al.) contain specific examples in which quantum mechanics and molecular dynamics are used to understand material performance as applied in the end-use application. As is typical when product development is the prime focus, the specific formulations cannot be divulged; but it is significant to learn how relative models may be applied to deduct material performance. Attention should be paid to the general characteristics and trend directions being modeled, rather than the specific identification of the material. Chapter 2 makes use of band structure and polarizability calculations to understand behavior of a dielectric material developed for flat panel displays, specifically when understanding voltage holding ratio properties during performance testing. While Chap. 2 concentrates on planewave techniques for electronic properties, Chap. 3 concentrates on molecular techniques for chemistry issues centering on the development of selective etchant/cleaner formulations. Since the etchant/cleaner material being examined is a formulation with a multitude of components, the modeling used in this chapter demonstrates a speculative approach in which the etch/clean performance must be inferred by the variety of molecular species present and its relative performance to others. Both quantum mechanics and Newtonian mechanical techniques are used to rate the etchant/cleaner, but because of the heavy use of quantum mechanics to understand the reaction thermodynamics of the complex mixture and the subtle impact of the molecular composition, the paper is included in this chapter.

## Part I Chapters List

**Chapter 1:** "Atomistic Simulations of Microelectronic Materials: Prediction of Mechanical, Thermal, and Electrical Properties"
V. Eyert, A. Mavromaras, D. Rigby, W. Wolf, M. Christensen, M. Halls, C. Freeman, P. Saxe and E. Wimmer

**Chapter 2:** "Using Molecular Modeling Trending to Understand Dielectric Susceptibility in Dielectrics for Display Applications"
Nancy Iwamoto, Ahila Krishnamoorthy and Edward W. Rutter Jr

**Chapter 3:** "Understanding Cleaner Efficiency for BARC ("Bottom Anti-Reflective Coating") After Plasma Etch in Dual Damascene Structures Through the Practical Use of Molecular Modeling Trends"
Nancy Iwamoto, Deborah Yellowaga, Amy Larson, Ben Palmer and Teri Baldwin-Hendricks

# Chapter 1
# Atomistic Simulations of Microelectronic Materials: Prediction of Mechanical, Thermal, and Electrical Properties

**V. Eyert, A. Mavromaras, D. Rigby, W. Wolf, M. Christensen, M. Halls, C. Freeman, P. Saxe and E. Wimmer**

This paper is based upon "Computational Materials Engineering: Capabilities of Atomic-Scale Prediction of Mechanical, Thermal, and Electrical Properties of Microelectronic Materials", by A. Mavromaras, D. Rigby, W. Wolf, M. Christensen, M. Halls, C. Freeman, P. Saxe, and E. Wimmer, which appeared in the Proceedings of Eurosime © 2010, IEEE.

**Abstract** The prediction of materials properties using atomic-scale simulations offers exciting and unprecedented opportunities to expand the capabilities of electronic devices, to create novel systems, and to improve their reliability. This contribution discusses the current state of atomic-scale simulations and their performance to predict mechanical, thermal, and electrical properties of microelectronic materials. Specific examples are the elastic moduli of compounds such as aluminum oxide, the strength of aluminum–silicon nitride interfaces, the coefficients of thermal expansion of bulk aluminum and silicon nitride, thermal conductivity of silicon and germanium, the prediction of the diffusion coefficient of hydrogen in metallic nickel, the calculation of dielectric properties of zinc oxide, and optical properties of silicon carbide and diamond. The final example addresses the control of the effective work function in the $HfO_2$/TiN interface of a CMOS gate stack. For an increasing number of materials properties, computed values achieve a level of accuracy which is similar to that of measured data. This enables the generation of consistent datasets of materials properties as basis for the design and optimization of materials in microelectronic devices. The insight and

V. Eyert (✉) · A. Mavromaras · D. Rigby · W. Wolf · M. Christensen
M. Halls · C. Freeman · P. Saxe · E. Wimmer
Materials Design, Inc, Santa Fe, NM 87501, USA
e-mail: veyert@materialsdesign.com

V. Eyert · A. Mavromaras · D. Rigby · W. Wolf · M. Christensen
M. Halls · C. Freeman · P. Saxe · E. Wimmer
Materials Design, SARL, 92120 Montrouge, France

N. Iwamoto et al. (eds.), *Molecular Modeling and Multiscaling Issues for Electronic Material Applications*, DOI: 10.1007/978-1-4614-1728-6_1,
© Springer Science+Business Media, LLC 2012

understanding gained by these simulations sets the stage for the development of innovative materials concepts, for example in the use of nanostructures and materials such as graphene.


## 1.1 Background

The capabilities currently used to understand and predict materials properties based on atomic-scale simulations have as their theoretical foundation four major theoretical concepts. These are classical mechanics as represented by Newton's laws of motion, the theory of electromagnetism as formulated in Maxwell's equations, statistical mechanics in the form pioneered mainly by Ludwig Boltzmann and Josiah Willard Gibbs, and quantum mechanics in the form of Schrödinger's equation including its relativistic generalization by Paul Dirac. During the middle of the twentieth century scientists such as Douglas Hartree and John Slater played a key role to develop practical approximations and computational methods to solve Schrödinger's equation for increasingly complex molecules and solids.

The development of density functional theory in the mid 1960s by Walter Kohn and coworkers combined with the breathtaking increase in computing power set the stage to apply new computational methods to materials of industrial relevance. The process of moving this field from academic research into the world of industrial laboratories started in the 1970s at places, such as the research laboratories of the IBM corporation in Yorktown Heights (New York), San Jose (California), and Rüschlikon (Switzerland), Bell Laboratories of then AT&T in Murray Hills (New Jersey), the Palo Alto Research Center of Xerox (California), the NEC Research Laboratory in Tsukuba (Japan), and others. However, it is fair to say that in the 1970s and 1980s the theoretical and computational work in these industrial laboratories had some fundamental research character, which was not dissimilar to academic research being pursued at that time in universities and at national laboratories.

This situation has now changed. Atomic-scale simulations are becoming an integral part of applied materials research and industrial engineering. One of the reasons for this transition is the revolution in the power of computers combined with their dramatic increase in cost-effectiveness. Furthermore, computational methods have matured and software systems have become available, which are designed for industrial productivity. At the same time, the growing interest in applying computational materials science to complex industrial problems exposes challenges which still have to be overcome, for example in dealing with the structural complexity of interfaces or in achieving the accuracy required in predicting reliable chemical reaction rates, and in modeling viscoelastic properties of polymers.

However, remarkable capabilities are available today for a range of materials and materials properties. This encompasses mechanical, thermal, electrical, optical, magnetic, and chemical properties of metals, semiconductors, ceramics, glasses, polymers, liquids, and gases. Many technological problems involve combinations of these materials, which makes simulations of interface phenomena a critical and challenging aspect of computational materials science.

A wide range of theoretical and computational approaches are being pursued, which are presently at quite different stages of maturity. These approaches include analytical theory, heuristic correlations such as quantitative structure–property relationships (QSPR), coarse-grain methods such as dissipative particle dynamics, molecular dynamics, and Monte Carlo methods employing interatomic empirical potentials (forcefields), semi-empirical quantum mechanical methods, and finally ab initio quantum mechanical methods.

Ab initio methods provide the highest predictive power, as no system-specific parameters are introduced in the simulations. However, this level of theory possesses three main limitations, namely the restriction to a relatively small system size (a few hundred atoms per simulation cell), the limitation to a very narrow region of phase space (a few thousand configurations), and the uncertainties due to approximations used in the solution of the many-body Schrödinger equation.

For these reasons, methods based on a quasi-classical description of interatomic interactions (often referred to as forcefield or empirical-potential methods) play an important role in predicting materials properties such as the density of a polymer and the thermal conductivity, which are derived from ensembles of hundreds and thousands of atoms and molecules and millions of different configurations.

The parameters required in forcefield simulations are either obtained purely by fitting experimental data (such as interatomic distances, elastic moduli, and vibrational properties) or by combining experimental data with information obtained from ab initio calculations on smaller systems. This creates a powerful combination of two complementary methods.

A large number of materials properties are now within reach of present computational methods as listed in Table 1.1.

## 1.2 Computational Approaches

### 1.2.1 Density Functional Theory

The most common quantum mechanical method for solid-state and molecular systems is based on density functional theory (DFT) as formulated in the mid 1960s by Hohenberg, Kohn, and Sham [1, 2]. This theory is based on the proven fact that in a system with many electrons such as a solid, the quantum-mechanical description of the N-electron system in its ground state can be exactly mapped onto N effective one-electron equations with an effective potential. The solutions of these so-called Kohn–Sham equations give rise to the electron density, which is identical to the electron density of the N-electron system. While being exact, application of DFT requires approximations regarding the effective potential.

**Table 1.1** Materials properties accessible by computational approaches: Q quantum mechanical, F forcefield methods

| **Structural properties** | |
|---|---|
| Bond distances and bond angles in molecules and nanoparticles | QF |
| Lattice parameters and atom positions of crystals | QF |
| Surface relaxations and reconstructions | QF |
| Defect structures | QF |
| Adsorption geometries | QF |
| Density and local structure of amorphous materials | QF |
| Density and local structure of polymeric materials | F |
| **Thermo-mechanical properties** | |
| Elastic moduli | QF |
| Cleavage energy | QF |
| Viscoelastic properties | F |
| Adhesion | QF |
| **Thermodynamic properties** | |
| Energy, enthalpy, entropy, free energy (Gibbs, Helmholtz) | QF |
| Heat capacity | QF |
| Solubility | QF |
| Melting temperature | F |
| Vapor pressure | F |
| Surface energy | QF |
| Interface energy | QF |
| **Chemical properties** | |
| Reaction rates in gas phase | Q |
| Surface reactivity, catalytic reactions, corrosion | Q |
| Photochemical reactions | Q |
| Electrochemical reactions | Q |
| **Transport properties** | |
| Mass diffusion | QF |
| Permeability | QF |
| Thermal conductivity | QF |
| Electrical conductivity | Q |
| Viscosity | F |
| **Electronic, optical, and magnetic properties** | |
| Dielectric properties | Q |
| Polarizability, hyperpolarizability | Q |
| UV and visible absorption spectra | Q |
| Frequency-dependent refractive index and reflectivity | Q |
| Piezoelectric properties | Q |
| Work function | Q |
| Energy band structure, band gaps | Q |
| Ionization energies, electron affinities | Q |
| Magnetic moments, magnetic anisotropy energy | Q |

In particular, it is assumed that the non-classical interactions between the electrons due to the effects of exchange and correlation can be described within a local picture, which takes into account only the electron density at the reference electrons position. This is called the "local density approximation (LDA)". If gradients of the electron density are included in the exchange–correlation terms, one obtains a group of approximations, which is referred to as "generalized gradient approximation" (GGA). A common form of a GGA is due to Perdew, Burke, and Ernzerhof (PBE) [3]. All electrostatic interactions between electrons as well as between electrons and nuclei can be treated without any approximations to the inhomogeneous shape of the charge distribution. With today's electronic structure programs such as the Vienna ab initio simulation package (VASP) [4, 5] it is now possible to model systems with hundreds of atoms per unit cell on a routine basis.

Many materials properties such as equilibrium structures, elastic coefficients, and vibrational frequencies (phonon dispersions), and also heats of formation and diffusion barriers are remarkably well described at the DFT-GGA level of theory. In fact, typical deviations between experimental and computed interatomic distances (and lattice parameters) are in the range of 1–2% and bond angles are described within 1–2° for a large variety of metals, semiconductors, insulators as well as organic molecules. Especially in the cases of point defects, surfaces, and buried interfaces this accuracy renders the method a powerful predictive tool, which rivals any experimental technique for measuring these structural properties.

However, this level of theory is not intended for the calculation of accurate excitation energies such as energy band gaps. In recent years, advanced methods such as hybrid functionals have been developed, which do provide very good agreement with experimental data for energy band gaps and related properties. These methods are computationally one to two orders of magnitude more demanding than DFT calculations at the GGA level. However, advances in algorithms and computer hardware are bringing these methods into the realm of routine materials engineering, as reviewed by Hafner [6].

## 1.2.2 Forcefield Methods

The simulation of many properties such as thermal conductivity, viscoelasticity, and vapor pressure involve a large number of particles and millions of configurations. For most of these systems quantum mechanical methods are too slow and also are often not accurate enough. In fact, present quantum-mechanical methods are very powerful to describe the formation and breaking of bonds and other strong interatomic interactions, but methods such as DFT-GGA fail to describe the subtleties of weak interactions between molecules, which determine properties such as interactions between entangled polymer chains. In such cases, one can describe the interatomic interactions by an analytic mathematical expression with adjustable parameters ("forcefield"). The choice of the form of such a forcefield is guided by chemical intuition and the parameters are fitted using a combination of experimental data and quantum-mechanical results obtained for small sub-systems or specific atomic arrangements.

The development of reliable forcefields is tedious and there is always a trade off to be made between accuracy and transferability. Organic molecules, polymers, and bio-macromolecules are particularly well suited for a forcefield approach while transferrable forcefields for inorganic materials have proved more difficult to obtain.

Prominent families of forcefields for organic and molecular systems include AMBER [7], CHARMM [8], Class 2 valence forcefields [9], OPLS [10], COMPASS [11], and the anisotropic united atom (AUA) potentials [12]. One of the simplest forms of an interatomic potential was introduced by Lennard-Jones [13], which is still often used for simple fluids. The Buckingham potential [14] is another commonly used potential for inorganic systems. In the case of ionic systems such as oxides it is common practice to use formal charges (e.g., +4 for silicon and −2 for oxygen) and compensate the resulting strong electrostatic interactions by appropriate repulsive terms in the potential.

The Stillinger–Weber potential [15] was originally designed to simulate elemental silicon. Another form of an empirical interatomic potential for inorganic materials was introduced by Tersoff [16]. Other potentials are those proposed by Finnis [17] and the bond-order potentials by Pettifor and Oleinik [18]. The "Gaussian Approximation Potentials" recently discussed by Csanyi and coworkers [19] present an interesting attempt to generalize the creation of interatomic potentials from ab initio calculations.

This very brief overview of forcefields and empirical potentials illustrates the long history and continuing evolution of this type of computational approach, which plays an important role in the practical applications of computer simulations. The main challenge of forcefield methods is to improve their predictive capability. Ultimately, a forcefield is limited by its functional form and the parameters used in its construction. In some cases, the accuracy of materials properties predicted with forcefield methods can be astounding and much better than the accuracy, which can be achieved with ab initio methods, but in other cases, without warning, the results can be disastrous. Thus, great care needs to be taken in the use of such methods. In fact, the judicious combination of ab initio methods and forcefield approaches does provide a way to increase the predictive power. This is one of the reasons why modern computational platforms emphasize the integration of different computational approaches in a single environment, facilitating the necessary cross-checking required to achieve confidence that predicted materials properties are of engineering value.

### 1.2.3 Other Computational Approaches

The third major class of computational approaches for the prediction of materials properties can be described as quantitative structure–property relationship (QSPR). In this type of approach one establishes correlations between descriptors and the properties of interest. By using chemical intuition and a sufficiently large training set one can derive rather remarkable correlations, which are very useful and quick. An example is the work by Bicerano on the prediction of polymer properties [20]. The recent work by Halls and Tasaki [21] on additives for lithium-ion batteries illustrates the application of correlation methods based on molecular descriptors derived from quantum mechanical calculations.

This short overview of computational methodologies illustrates the breadth and variety of approaches, which can be used for the prediction of materials properties of interest to the semiconductor industry. We will now discuss several specific examples, which demonstrate current capabilities with a focus on the quantitative prediction of materials properties using ab initio methods.

## 1.3 Examples

### 1.3.1 Elastic Coefficients of Aluminum Oxide

The elastic coefficients of a material are of fundamental importance as they describe the response of a solid to external forces in the regime of small displacements (the elastic regime). Once the elastic coefficients are known, one can readily compute the elastic moduli (bulk, Young, and shear modulus) as well as related properties such as the speed of sound within the solid.

Despite their importance, the total number of accurately measured elastic coefficients is relatively small (approximately a few thousand) and there is substantial scatter in the experimental data. This is due to the fact that a precise experimental determination of elastic coefficients requires a high-purity single crystal and rather subtle experimental procedures.

Corundum ($\alpha$-$Al_2O_3$) is an intriguing example of the capabilities of today's quantum-mechanical methods. The computed elastic coefficients are in good agreement with earlier experiments (cf. Expt.[a] in Table 1.2) except that the coefficient $C_{14}$ has the opposite sign compared with the computed value. An in-depth analysis of this discrepancy revealed that since 1960, all the literature results of the elastic tensor of corundum ($\alpha$-$Al_2O_3$) correspond to an unintended reverse setting of the rhombohedral lattice in the hexagonal axes. This leads to a change in the sign of the elastic coefficient $C_{14}$, as seen in Expt.[a] in Table 1.2. Computations of the elastic coefficients revealed this mistake and new measurements by Gladden et al. [23] corrected this inconsistency also on the experimental side.

### 1.3.2 Strength of Al/Si$_3$N$_4$ Interface

The study of interfaces is one specific area where computational materials research can provide information not easily obtained by experimental methods.

**Table 1.2** Computed and experimental elastic coefficients of corundum, $\alpha$-Al$_2$O$_3$

|          | Expt.[a] | Expt.[b] | | Computed.[b] |
|----------|----------|----------|----------|----------|
|          |          | Sample 1 | Sample 2 |          |
| $C_{11}$ | 497.3    | 495.6    | 497.4    | 495      |
| $C_{12}$ | 162.8    | 160.2    | 158.3    | 171      |
| $C_{13}$ | 116.0    | 117.0    | 121.0    | 130      |
| $C_{14}$ | **−21.9**| +22.1    | +23.0    | +20      |
| $C_{33}$ | 500.9    | 501.0    | 505.8    | 486      |
| $C_{44}$ | 146.8    | 147.0    | 145.3    | 148      |

[a] Ref. [22]
[b] Ref. [23]
All values in GPa

**Fig. 1.1** Computed equilibrium structure and work of separation of an Al/Si$_3$N$_4$ interface. Note the pronounced difference depending on the termination of the ceramic phase

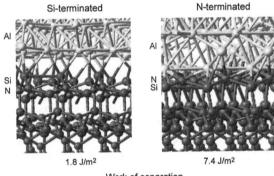

Many applications are critically dependent on strong bonding at the contact points between different materials or phases of materials. This, for example, is the case for many metal-ceramic systems, such as aluminum–silicon nitride, where two very dissimilar materials are combined (Fig. 1.1).

Silicon nitride, Si$_3$N$_4$, is a covalently bonded ceramic with desirable mechanical properties such as high strength, corrosion resistance, and good thermal shock resistance. However, many applications are hindered by the brittleness of the compound. By mixing silicon nitride with aluminum, the toughness can be increased (and the composite can even be superplastic). Joining aluminum and silicon nitride requires a good adhesion between the ceramic and the metal. Computational methods, in particular DFT, are well suited to assess the energetics of such interfaces.

A model of the Al/Si$_3$N$_4$ interface is constructed by matching the (001) surface of Al with the (100) surface of $\beta$-Si$_3$N$_4$ (cf. Fig. 1.1). The silicon nitride surface can be either fully silicon terminated or nitrogen terminated. A systematic search of possible interfaces with a good geometric match between the two surfaces leads to a computational model consisting of 134 atoms for the Si-terminated silicon nitride and 122 atoms for the N terminated nitride. The interface area in the models is 67.23 Å$^2$ with a mismatch between the two lattices of 1.9 and −3.2% of the in-plane lattice parameters and less than 1° in the lattice angles. All atom positions are relaxed using a conjugate gradient method with ab initio forces computed with VASP [4, 5] in the MedeA environment [24].

The strength of the interface is assessed by computing the ideal work of separation. This is the work required to separate the two phases forming the interface into two non-interacting free surfaces. This computed quantity simplifies the cleavage by ignoring dissipative and plastic processes, but it gives a well-defined and computationally accessible assessment of the interface strength (i.e., the limit of brittle fracture).

The computational results show that the interface strength vitally depends on the details of the atomic arrangement directly at the $Al/Si_3N_4$ interface. Aluminum forms much stronger interatomic bonds with nitrogen than with silicon at the interface. A completely nitrogen terminated interface is about four times as strong as a completely silicon terminated interface (with a work of separation of 7.4 vs. 1.8 $J/m^2$) as illustrated in Fig. 1.1. Hence, although the free (100) surface of $Si_3N_4$ is more stable with Si as the outermost layer, this stability is reversed at the interface. Any production process aiming at good adhesion should therefore consider the nitrogen chemical potential to facilitate N termination at the interface.

## *1.3.3 Thermal Expansion*

A remarkable capability of ab initio methods is the calculation of the coefficients of thermal expansion. This is possible by computing the complete vibrational spectrum (the so-called phonon dispersions) for a series of different lattice parameters. Knowledge of the phonon dispersions provides the phonon density of states, i.e., the number of lattice vibrational modes as a function of frequency. The phonon density of states contains all information necessary to compute the vibrational free energy (enthalpy and entropy) as a function of temperature. The free energy then determines the lattice parameters as a function of temperature, i.e., the thermal expansion of a solid.

This procedure is demonstrated here for the two components of the interface discussed in the previous section, namely bulk Al and $Si_3N_4$. As expected, the calculations reveal a high thermal expansion of metallic aluminum whereas the ceramic material $Si_3N_4$ has a very small coefficient of thermal expansion, as shown in Fig. 1.2.

The ab initio calculations are carried out on the DFT-GGA level of theory with the PBE potentials. The all-electron frozen core projector augmented plane-wave method is used as implemented in VASP [4, 5] with an energy cutoff of 500 eV. A total of 10 phonon calculations are performed for Al and 15 for silicon nitride for a range of lattice parameters to obtain enough reference points for the free energy. The phonon calculations are performed within the direct approach as formulated by Parlinski [25] and implemented in the MedeA software environment [24].

**Fig. 1.2** Computed temperature-dependent lattice parameters of bulk Al and $Si_3N_4$. The ceramic material has a hexagonal lattice and the thermal expansion in the $c$-axis is slightly smaller than in the basal plane. The *lines* are drawn up to the melting points of 933.47 K for Al and 2,173 K for $Si_3N_4$

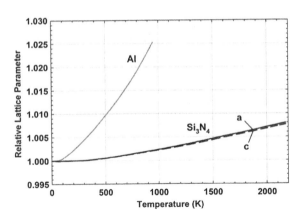

### 1.3.4 Diffusion of Atoms in Solids

Diffusion of atoms in solids is a phenomenon of broad industrial interest both in the manufacturing of devices as well as in predicting the life-time behavior of systems. Despite this importance, reliable data for diffusion coefficients are actually quite sparse. For example, experimental values for the diffusion coefficient of hydrogen atoms in metals such as titanium and aluminum are scattered over many orders of magnitude and values for many important systems are missing.

One of the systems where a large number of experimental data have lead to a consistent picture of diffusion is hydrogen in nickel. For this reason an in-depth computational study has been performed to probe the accuracy of an ab initio approach for computing the mass diffusion coefficients [26].

The temperature-dependent diffusion coefficient, $D$, is conveniently written as $D = D_0 \exp[-Q/(RT)]$ with $Q$ being the diffusion barrier, $R$ the gas constant, $T$ the absolute temperature, and $D_0$ the pre-exponential factor. The computation is based on the following model and theoretical concepts. The solubility of H in Ni is relatively low, so we consider a dilute system where the interaction between different H atoms can be neglected. The system is modeled by a periodic simulation cell containing 32 Ni atoms and one hydrogen atom. The H atom is placed at different interstitial sites and all atoms of the system are relaxed until the total energy of the system reaches a minimum. The total-energy calculations are performed on the DFT-GGA level of theory using the VASP program [4, 5] as integrated in the MedeA environment [24]. Magnetic effects are taken into account by using a spin-polarized effective potential in the Kohn–Sham equations. The calculations show that in their most stable state hydrogen atoms reside in octahedral interstitial sites. The energetically most favorable diffusion path leads from the octahedral site across a transition state into a metastable tetrahedral site and onward to another adjacent octahedral site.

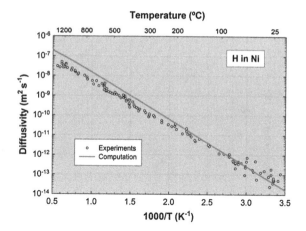

**Fig. 1.3** Computed and experimental diffusion coefficient for hydrogen atoms in metallic nickel. After Ref. [26]

The temperature-dependent jump rate is obtained with Eyring's transition state theory [27]. In this approach the probability of a jump to occur depends on the height of the diffusion barrier and on the vibrational frequencies both at the stable ground state and at the transition state. To this end, all vibrational modes of the 33-atom model are computed for H at the stable octahedral interstitial site, the transition state, and the metastable tetrahedral site using the direct method for phonon calculations. Thermal expansion of the Ni lattice is taken into account by using a coefficient of thermal expansion, which is also computed from first-principles phonon calculations as described in the previous section. The resulting temperature-dependent diffusion coefficient is conveniently represented in the form of an Arrhenius plot as shown in Fig. 1.3. The slope of the curve $D(T)$ is related to the effective height of the diffusion barrier and the absolute position on the $y$-axis is defined by the pre-factor $D_0$.

The resulting agreement between computed and experimental data at temperatures up to about 200°C is remarkable given that the calculations are solely obtained from fundamental physical constants, the laws of quantum mechanics, and statistical mechanics. The deviation of experimental and computed values at elevated temperatures may be due to several factors including the presence of defects in the samples acting as hydrogen traps as well as limitations of the quasi-harmonic approximation, which is known to be less appropriate for elevated temperatures.

In view of the lack of reliable and systematic experimental data for diffusion coefficients as well as the scatter in some of the available data, the present result indicates that computed diffusion coefficients can reach a level of accuracy which is comparable with experiment. Thus, computations are a valuable source of diffusion data.

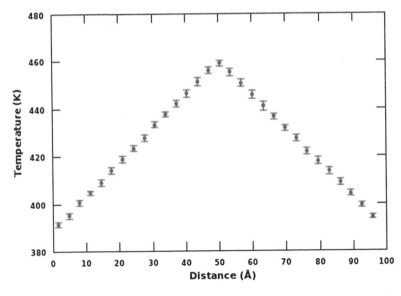

**Fig. 1.4** Computed temperature profile used for determining the thermal conductivity of hexane using non-equilibrium molecular dynamics

### 1.3.5 Thermal Conductivity

The computation of thermal conductivity can be achieved by analyzing the fluctuations in a molecular dynamics simulations using the so-called Green–Kubo method or, alternatively, using non-equilibrium molecular dynamics simulations. A particularly elegant means of accomplishing this task is a method developed by Müller-Plathe et al. [28]. In this approach, the system is modeled by an elongated supercell. The cell is divided into slices perpendicular to the long axis of the model and molecular dynamics simulations are performed such that the middle regions become hot and the outer regions remain cold.

This is achieved by swapping approximately every 100–1,000 steps the hottest particle (highest kinetic energy) from the "cold" region and the coldest particle from the "hot" region. After a sufficiently long simulation time a temperature gradient builds up and a quasi-stationary state is reached. From the temperature gradient and the heat flux (which is defined by the swapping of the momenta) one obtains the thermal conductivity of the system. The illustrative calculation for hexane, shown in Fig. 1.4, uses a Class 2 forcefield and the LAMMPS program [29] within the MedeA platform [24].

Another example of the ability to compute thermal conductivity is the case of solid Si and Ge. By using a Green–Kubo approach and the COMPASS forcefield, the calculations result in a value of 272 W/(mK) for crystalline silicon (see Table 1.3). This is more than a factor of two higher than experiment. The difference originates from an isotope effect. Natural silicon is a mixture of isotopes with a

**Table 1.3** Computed and experimental thermal conductivity of crystalline Si and Ge

|     | Isotopically pure | Isotopic mixture | Expt. |
| --- | --- | --- | --- |
| Si  | 272 | 128 | 130 |
| Ge  | 98  | 37  | 58  |

Note the pronounced isotope effect. The experimental values are from the web site of the Ioffe Institute, http://www.ioffe.rssi.ru/ SVA/NSM/Semicond/SiGe/thermal.html

representative composition of 92.2% $^{28}$Si, 4.7% $^{29}$Si, and 3.1% $^{30}$Si. It is quite remarkable that about 8% admixture of slightly heavier nuclei has such a dramatic effect on the thermal conductivity. A similar effect is seen for germanium.

## 1.3.6 Band Gaps

For a number of decades density functional calculations for semiconductors were plagued by the "band gap problem", namely the fact that standard DFT calculations employing the usual approximations such as LDA and GGA result in band gaps, which are typically 30–50% too small as compared with experimental values. This result is understandable because this level of theory is intended to predict ground state properties and not electronic excitation energies. In some cases such as elemental Ge and InSb, standard DFT calculations do not predict any gap at all. Today, this difficulty has been overcome by methods, which go beyond DFT. Hybrid functionals such as those proposed by Heyd et al. [30] offer a pragmatic solution. This is illustrated here for the case of Ge.

As can be seen from Fig. 1.5, the computed and experimental band gap and features of the energy band structure agree very well. In particular, while standard DFT calculations using the LDA or GGA give very tiny or even vanishing band gaps, hybrid functional calculations result in a perfect agreement between the measured and the calculated value of 0.66 eV.

## 1.3.7 Dielectric Properties

The computed dielectric properties of hexagonal ZnO are an excellent example of the performance of present day electronic structure calculations, as demonstrated recently by Wrobel et al. [31].

The calculations were performed with VASP 5.2 using the recent implementation of the hybrid functional of Heyd-Scuseria-Ernzerhof (HSE) [30] as was done for the energy band structure of Ge. The dielectric constant $\varepsilon^0$ corresponds to the low-frequency limit, where the ions are fully relaxed. In the high-frequency limit, the dielectric tensor is smaller since the ions do not have time to relax and only the response of the electron system contributes to this property.

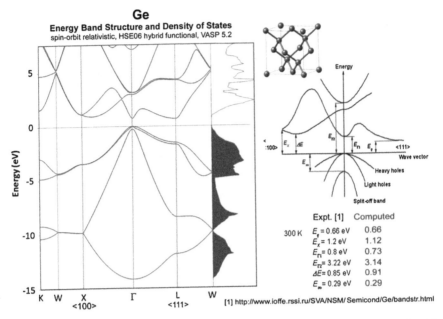

**Fig. 1.5** Computed energy band structure of elemental Ge (*left panel*) compared with experimental data

**Table 1.4** Computed and experimental dielectric properties of hexagonal ZnO in units of $\varepsilon_0$. After Ref. [31]

| | Direction | Computed | Expt. [32] |
|---|---|---|---|
| $\varepsilon^\infty$ | 11 | 3.66 | 3.70 |
| $\varepsilon^\infty$ | 33 | 3.73 | 3.78 |
| $\varepsilon^0$ | 11 | 7.72 | 7.77 |
| $\varepsilon^0$ | 33 | 8.37 | 8.91 |

It can be seen from Table 1.4 that the agreement of computed and experimental data is very good. This opens the possibility to investigate dielectric properties for a material not just at ambient conditions, but, for example, also at high pressures or at anisotropic stress. Given the predictive power of ab initio methods, the predicted materials properties are likely to be similar in quality to the values obtained for ambient conditions, where a direct comparison with experiment is possible.

### 1.3.8 Optical Properties

As illustration of the capabilities to compute optical properties we present two examples, namely (1) the reflectivity of silicon carbide in the cubic and the hexagonal 6*H* polymorphs and (2) the refractive index of diamond. In both cases the

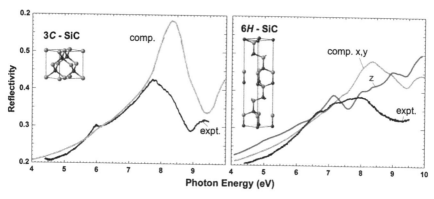

**Fig. 1.6** Computed and experimental [33] reflectivity of cubic and hexagonal (6*H*) silicon carbide

calculations were performed at the ab initio level of theory with VASP 5.2 using the Heyd-Scuseria-Ernzerhof hybrid functional [30] within the MedeA environment.

The computed reflectivity of SiC agrees fairly well with experiment [33] up to about 7 eV as shown in Fig. 1.6. For example, the small peak at 6 eV seen in the experiment for the cubic form (3*C*) is also found in the calculations as a shoulder. At higher energies the experimental maxima near 7.7 eV (3*C*) and 8.0 eV (6*H*) are shifted to higher energies in the computations. It is likely that this deviation is caused by the specific form of the non-local exchange–correlation potential. The Hartree–Fock-like component of this potential tends to shift unoccupied eigenvalues to higher energies. Possibly, this shift is too strong in the present case. In fact, standard DFT-LDA or DFT-GGA calculations would shift this peak to much lower energies. Given the theoretical challenges inherent in the computation of excitation energies in solids, the present results are reasonable and encouraging, although further improvements are certainly desirable.

The second example of computed optical properties is the refractive index of diamond as a function of energy as illustrated in Fig. 1.7. The computed curve tracks the experimental data points beautifully. The computed values are systematically slightly smaller than experiment. This may be related to the fact that the HSE hybrid functional overestimates the band separation, which was also seen in the case of SiC. Nevertheless, the computed result is quite impressive given that it was obtained from an ab initio quantum-mechanical approach, which has no system-specific parameters.

## 1.3.9 Effective Work Function in CMOS Gate Stack

The last example is related to the introduction of hafnium oxide as a high-k dielectric in complementary metal–oxide–semiconductor (CMOS) devices.

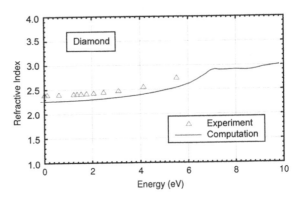

**Fig. 1.7** Computed and experimental refractive index of diamond

The use of oxides other than $SiO_2$ necessitates a change of the metallic gate material. The effective work function (EWF) of the metal must be tuned such that for n-doped devices (NMOS) the Fermi energy of the metal gate lines up closely with the bottom of the conduction band of the Si channel while for PMOS the Fermi level should be close to the valence band maximum. In other words the EWF needs to be tuned over the range of the band gap in Si with a high EWF for PMOS and a low EWF for NMOS devices.

Titanium nitride is being considered as one of the possible metallic gate materials. The tuning of EWF has posed considerable challenges [34] and the understanding of the important effects is still developing despite substantial investigation.

In this context ab initio simulations provide unique insight into the relationship between chemical composition, structure, and work function. For such an analysis the key quantity is the electrostatic potential across the gate stack in the direction perpendicular to the interfaces. Similar to work function changes at surfaces, the formation of dipole layers due to charge transfer at the metal/oxide interface are critical in determining the EWF. For this reason, the $HfO_2$/TiN interface and the role of the chemical composition and structure on interface dipole layers are of particular interest.

A model of the $HfO_2$/TiN interface (Fig. 1.8) is constructed as follows. Hafnium oxide is represented by a film of crystalline monoclinic $HfO_2$ with its most stable surface forming the interface. Despite the partially amorphous character of the oxide film in actual systems, a crystalline model is reasonable since the coordination of Hf in amorphous and annealed oxides is likely to reflect the key local structural aspects such as nearest-neighbor bond distances and bond angles as found in stable crystalline structures. A series of calculations on all low-index surfaces of monoclinic $HfO_2$ shows that the stoichiometric $(\bar{1}11)$ surface has the lowest surface energy. This surface is taken as substrate.

A systematic search for a good geometric match between the $HfO_2(\bar{1}11)$ surface and a N-terminated TiN(111) surface leads to a computationally convenient supercell containing 12 Hf atoms and 16 Ti atoms per layer. The interface area in this supercell is 127.37 $\mathring{A}^2$ with a mismatch between the two lattices of

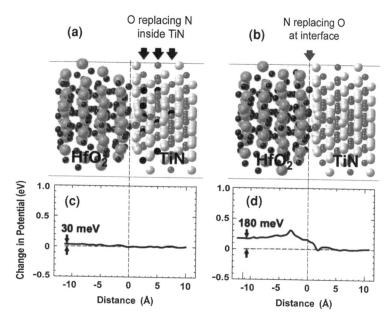

**Fig. 1.8** Changes of the electrostatic potential across a HfO$_2$/TiN interface computed for (*left*) a case when nitrogen atoms inside TiN are replaced by oxygen atoms and (*right*) when O atoms are replaced by N atoms at the metal-oxide interface. After Ref. [34]

only 0.5 and $-1.9\%$ of the in-plane lattice parameters and less than $4°$ in the lattice angles. Using this geometry first a few layers of TiN in the sequence Ti–N–Ti–N are deposited on the oxide surface and the system is relaxed by ab initio molecular dynamics simulated annealing. Subsequently additional layers of Ti and N with a capping layer of Ti are added, which results in a supercell containing 288 atoms of the composition Hf$_{48}$O$_{96}$Ti$_{80}$N$_{64}$.

All atom positions of the system are relaxed using a conjugate gradient method with ab initio forces computed with VASP [4, 5] in the MedeA environment [24]. The electrostatic potential of this model is averaged in planes parallel to the interface, and smoothed using a macroscopic average [35]. This establishes a reference for the electrostatic potential across the interface.

Experimentally it has been found that annealing of HfO$_2$/TiN stacks under an oxygen-containing atmosphere leads to incorporation of oxygen atoms in the TiN film and an increase of the EWF by up to 550 meV [34]. Hence the first question to answer is the effect on the EWF when N atoms are replaced by O in the TiN. Specifically, 14% of the N atoms in the TiN film are replaced by O, all atom positions are relaxed and the resulting planar-averaged electrostatic potential is compared with that of the reference system. As illustrated in Fig. 1.8, this modification of TiN has very little effect on the EWF and thus cannot be the cause of the observed work function increase.

Calculations reveal that modifications directly at the $HfO_2$/TiN interface can have a major impact. This is illustrated by replacing 1/3 of the O atoms at the interface by N atoms (see panels b and d in Fig. 1.8). This modification causes an increase of 180 meV in the EWF. At first this appears to be counter-intuitive. One might expect that the replacement of strongly electronegative atoms such as O by the less electronegative N atoms would lead to a decrease in the work function. This would be true for a surface of TiN, but not an interface, where one also has to consider the electronic rearrangement around Hf. In fact the replacement of O by N at the interface leads to changes of two opposing dipoles, namely that of Hf-X and X-Ti (X = O and N). Hf atoms are larger and more polarizable than Ti atoms, so the changes in the Hf-X layers prevail and the net effect is an increase of the work function viewed from the TiN side of the stack.

This insight gained from the ab initio calculations of these gate-stack materials provides guidelines for the optimization of the process conditions in the fabrication of these stacks. The control of the chemical composition at the interface at the scale of a single atomic layer is of critical importance while the EWF is rather insensitive to changes inside the TiN layer. This study also demonstrates that ab initio calculations can now be performed on models, which are sufficiently large to capture main features of actual devices such as metal-oxide interfaces.


## 1.4 Summary and Conclusions

Atomic-scale computational materials engineering is becoming an industrial practice. This achievement is based on the steady progress of theoretical and computational methods combined with the dramatic increase in compute power and cost-effectiveness. These developments have fueled the creation of integrated computational software systems, which are designed as productivity tools for industrial scientists and engineers.

One of the most exciting achievements is the ability to use ab initio quantum-mechanical methods for predicting a range of important materials properties of industrial relevance at an unprecedented level of detail. This is the major focus of the present article.

The illustrative examples start with the calculation of elastic coefficients of alumina, $Al_2O_3$, where the computations revealed an inconsistency in the experimental values, which had crept into the literature in the 1960s and has remained undetected until ab initio calculations provided clear evidence of this problem and helped to resolve it. In other words, one can say that due to their systematic performance, ab initio calculations can help to improve the consistency of experimental data. Furthermore, it is much faster to compute elastic coefficients than to synthesize single crystals and perform subtle measurements. In this case, as in others, the reliability of computed results is comparable to the reliability of experimental data. Within industrial research and development programs such cost-effectiveness and reliability in computed properties translates into significant gains in time and cost.

The second example of the strength of an $Al/Si_3N_4$ interface demonstrates that computations can provide detailed insight into bonding mechanisms, which are not readily accessible by experiment. This example also shows a limitation. Only one of many possible interface structures has been investigated and it is difficult to judge the relevance of the model. While the calculations are consistent and accurate, the uncertainty lies in the formulation of the question. Hence, close cooperation between experimental and computational efforts is needed to derive the highest value from the simulations.

The third example demonstrates the strengths of ab initio calculations of thermal expansion. For a long time, electronic structure theory was condemned to the "$T = 0$ prison". This limitation has been overcome. Using a quasi-harmonic phonon approach as was done here or directly using ab initio molecular dynamics, temperature-dependent materials properties have now come within the reach of ab initio methods.

The subsequent case shows that temperature-dependent diffusion coefficients can be obtained by ab initio calculations. In fact, the computed diffusion coefficient for hydrogen atoms in metallic nickel is in remarkable agreement with experimental data. This opens the possibility of investigating diffusivity for systems where no experimental data are available or for conditions such as regions of high stress, for which experiments are very difficult or impractical. Again a word of caution needs to be stated. Finding the correct diffusion mechanism is by no means a clear-cut procedure. Premature assumptions about a diffusion mechanism may lead to misleading conclusions, even if the calculations are done correctly. Again, a close collaboration between experimentalists and computational experts is called for.

In the realm of transport properties thermal conductivity is of great technological importance, especially in nanostructured microelectronic materials. The example shown in this article illustrates that this property can be computed using quasi-classical forcefields. In the case of Si and Ge, the calculations reveal a pronounced isotope effect, which illustrates the high sensitivity of the present computational approach to variations in the masses in these semiconducting materials.

As discussed in the introduction, there are a large number of different classes of forcefields and the determination of parameters is a subtle process. Before thermal conductivity can be computed for any kind of material, substantial efforts have to be made in the assessment of the appropriate forcefields and parameters. Again, the availability of large computing capacity and highly automated software systems will allow this task to be carried out efficiently.

The subsequent example demonstrates the capabilities of advanced electronic structure approaches (i.e., post-DFT methods) to compute energy band structures and dielectric properties. In the case of elemental Ge, the computations show excellent agreement with experimental energy band gaps and features of the band structure. This capability represents a major step forward in the usefulness of electronic structure calculations for semiconducting materials. An accurate computational description of the energy band structure is a prerequisite for the reliable calculation of dielectric properties. In fact, the results for hexagonal ZnO are very encouraging and it can be expected that this capability will find widespread use in the future.

The prediction of optical spectra is illustrated for the cubic and a hexagonal polymorph of silicon carbide and the refractive index of diamond. The results for SiC are encouraging, though more work is needed to achieve fully quantitative agreement between experiment and calculations. Possibly, there are also effects present in the experimental data, which cloud a direct comparison with computed results for ideal bulk systems. The computed refractive index of diamond is in remarkable agreement with experiment, thus opening the possibility to design optical materials with specific refractive properties prior to their actual synthesis.

The final example shows the impact of detailed ab initio calculations on the understanding of the factors, which control the work function in a $HfO_2$/TiN interface. In this case, the calculations point to the critical importance of the composition of a single layer at the metal/oxide interface, thus helping to focus experimental efforts.

In summary, the predictions of materials properties using atomic-scale simulations and electronic structure theory have advanced to a stage of industrial acceptance and integration. It is fair to say that we are still at the beginning of this exciting development. Persistence and hard work will be required both on the fundamental research level as well as in the transformation of research approaches into engineering tools. The creation of better materials and better processes for our societies, which computational materials engineering increasingly contributes to, certainly justifies this effort.

**Acknowledgments** The authors are indebted to many Materials Design partners for vital and stimulating discussions leading to the work summarized here. In particular the contributions of Dr. Jim Chambers, Dr. Hiro Niimi, and Judy Shaw at Texas Instruments, Prof. Jürgen Hafner and Prof. Georg Kresse at the University of Vienna, and Prof. Chris Hinkle at the University of Texas, Dallas are gratefully acknowledged.

# References

1. Hohenberg P, Kohn W (1964) Inhomogeneous electron gas. Phys Rev 136:B864
2. Kohn W, Sham LJ (1965) Self-consistent equations including exchange and correlation effects. Phys Rev 140:A1133
3. Perdew JP, Burke K, Ernzerhof M (1996) Generalized gradient approximation made simple. Phys Rev Lett 77:3865
4. Kresse G, Hafner J (1993) Ab initio molecular dynamics for liquid metals. Phys Rev B 47:558
5. Kresse G, Furthmüller J (1996) Efficient iterative schemes for ab initio total-energy calculations using a plane-wave basis set. Phys Rev B 54:11169
6. Hafner J (2008) Ab initio simulations of materials using VASP: density functional theory and beyond. J Comput Chem 29:2044

7. Cornell WD, Cieplak P, Bayly CI, Gould IR, Merz KM Jr, Ferguson DM, Spellmeyer DC, Fox T, Caldwell JW, Kollman PA (1995) A second generation force field for the simulation of proteins, nucleic acids, and organic molecules. J Am Chem Soc 117:5179
8. Brooks BR, Bruccoleri RE, Olafson BD, States DJ, Swaminathan S, Karplus M (1983) CHARMM: a program for macromolecular energy, minimization, and dynamics calculations. J Comp Chem 4:187
9. Maple JR, Dinur U, Hagler AT (1988) Derivation of force fields for molecular mechanics and dynamics from ab initio energy surfaces. Proc Natl Acad Sci 85:5350
10. Jorgensen WL, Maxwell DS, Tirado-Rives J (1996) Development and testing of the OPLS all-atom force field on conformational energetics and properties of organic liquids. J Am Chem Soc 188:1125
11. Sun H (1998) COMPASS: an ab initio force-field optimizied for condensed-phase applications—overview with details on alkane and benzene compounds. J Phys Chem B 102:7338
12. Ungerer P, Beauvais C, Delhommelle J, Boutin A, Fuchs A (2000) Optimization of the anisotropic united atoms intermolecular potential for n-alkanes. J Chem Phys 112:5499
13. Jones JE (1924) On the determination of molecular fields II. From the equation of state of a gas. Proc Roy Soc 106:463
14. Buckingham RA (1938) The Classical equation of state of gaseous helium, neon and argon. Proc Roy Soc London Ser A 168:264
15. Stillinger FH, Weber TA (1985) Computer simulation of local order in condensed phases of silicon. Phys Rev B 31:5262
16. Tersoff J (1988) New empirical approach for the structure and energy of covalent systems. Phys Rev B 37:6991
17. Finnis MW (1984) A simple empirical N-body potential for transition metals. Phil Mag A 50:45
18. Pettifor DG, Oleinik II (1999) Analytic bond-order potentials beyond Tersoff-Brenner. I. Theory. Phys Rev B 59:8487
19. Bartók AP, Payne MC, Kondor R, Csányi G (2010) Gaussian approximation potentials: the accuracy of quantum mechanics without the electrons. Phys Rev Lett 104:136403
20. Bicerano J (1993) Prediction of polymer properties. Marcel Dekker Inc, New York
21. Halls MD, Tasaki K (2010) High-throughput quantum chemistry and virtual screening for lithium ion battery electrolyte additives. J Power Sources 195:1472
22. Goto T, Anderson OL, Ohno I, Yamamoto S (1989) Elastic constants of corundum up to 1825 K. J Geophys Res 94:7588
23. Gladden JR, Maynard JD, So JH, Saxe P, Le Page Y (2004) Reconciliation of ab initio theory and experimental elastic properties of $Al_2O_3$. Appl Phys Lett 85:392
24. Materials Design, Inc (2010) MedeA 2.5, Santa Fe, NM
25. Parlinski K, Li Z Q, and Kawazoe Y (1997) First-principles determination of the soft mode in cubic $ZrO_2$. Phys Rev Lett 78:4063 and references therein
26. Wimmer E, Wolf W, Sticht J, Saxe P, Geller CB, Najafabadi R, Young GA (2008) Temperature-dependent diffusion coefficients from ab initio computations: hydrogen, deuterium, and tritium in nickel. Phys Rev B 77:134305
27. Eyring H (1935) The activated complex in chemical reactions. J Chem Phys 3:107
28. Müller-Plathe F, Reith D (1999) Cause and effect reversed in non-equilibrium molecular dynamics: an easy route to transport coefficients, Comp Theor Polymer Sci 9:203
29. Plimpton SJ (1995) Fast parallel algorithms for short-range molecular dynamics. J Comp Phys 117:1. http://lammps.sandia.gov
30. Heyd J, Scuseria GE, Ernzerhof M (2003) Hybrid functionals based on a screened Coulomb potential. J Chem Phys 118:8207
31. Wrobel J, Kurzydlowski KJ, Hummer K, Kresse G, Piechota J (2009) Calculations of ZnO properties using the Heyd-Scuseria-Ernzerhof screened hybrid density functional. Phys Rev B 80:155124

32. Ashkenov N, Mbenkum BN, Bundesmann C, Riede V, Lorenz M, Spemann D, Kaidashev EM, Kasic A, Schubert M, Grundmann M, Wagner G, Neumann H, Darakchieva V, Arwin H, Monemar B (2003) Infrared dielectric functions and phonon modes of high-quality ZnO films. J. Appl. Phys. 93:126
33. Lambrecht WRL, Segall B, Yoganathan M, Suttrop W, Devaty RP, Choyke WJ, Edmond JA, Powell JA, Alouani M (1994) Calculated and measured uv reflectivity of SiC polytypes. Phys Rev B 50:10722
34. Hinkle CL, Galatage RV, Chapman RA, Vogel EM, Alshareef HN, Freeman C, Wimmer E, Niimi H, Li-Fatou A, Shaw JB, Chambers JJ (2010) Interfacial oxygen and nitrogen induced dipole formation and vacancy passivation for increased effective work functions in TiN/HfO$_2$ gate stacks. Appl Phys Lett 96:103502
35. Peressi M, Baroni S, Baldereschi A, Resta R (1990) Electronic structure of InP/Ga$_{0.47}$In$_{0.53}$As interfaces. Phys Rev B 41:12106

# Chapter 2
# Using Molecular Modeling Trending to Understand Dielectric Susceptibility in Dielectrics for Display Applications

**Nancy Iwamoto, Ahila Krishnamoorthy and Edward W. Rutter Jr**

This paper is based upon "Understanding Leakage Current Susceptibility in Dielectrics using Molecular Modeling", which appeared in the Proceedings of Eurosime © 2009, IEEE.

**Abstract** Dielectric materials are universally used in the fabrication and packaging of microelectronic and display devices, as well as used in discrete devices such as sensors, switches and photovoltaics. However, as device and interconnect sizes become smaller, the question of the source of electrical failure becomes more and more important. During the development of these materials it has been found that small changes in the molecular structure can lead to small increases in conductivity which is undesirable for most applications. Although important to the final commercial acceptance of the dielectric, leakage current is often one of the final properties measured when developing the chemistry of the dielectric, so a dielectric with very good mechanical properties can ultimately fail at end-user applications due to the poor electrical properties. Knowledge of the susceptibility for electrical failure can be a great aid to the developer, and molecular modeling used in a trend analysis has been found useful to predict tendencies.

## 2.1 Introduction and Problem Focus

This paper focuses on the use of molecular modeling to help explain experimental observations during the development of thick film dielectrics for application in flat panel displays (FPD). The layer configurations used in these displays are

N. Iwamoto (✉) · A. Krishnamoorthy · E. W. Rutter Jr
Honeywell Specialty Materials, 1349 Moffett Park Drive,
Sunnyvale, CA 94089, USA
e-mail: nancy.iwamoto@honeywell.com

N. Iwamoto et al. (eds.), *Molecular Modeling and Multiscaling Issues for Electronic Material Applications*, DOI: 10.1007/978-1-4614-1728-6_2, © Springer Science+Business Media, LLC 2012

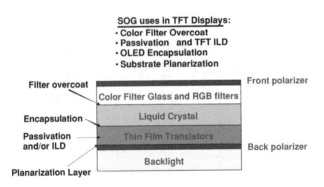

**Fig. 2.1** General layer build-up in a display

well-known [1] and a schematic is found in Fig. 2.1. The traditional spin-on-glass (SOG) formulation may have an application in many of the different layers used, including uses such as planarization, passivation and encapsulant layers, however the materials under development in the current work were used as encapsulant layers next to the liquid crystal layer.

As may be envisioned in such a multi-layer configuration, if materials are to be correctly integrated, the electrical effects on the total performance of the display should be understood. Since the target material application involved the encapsulation layer next to the liquid crystal, it was important to show that the dielectric under development would not interfere with the switching of the liquid crystal, and molecular modeling was found to be very useful in that understanding.

The electrical properties of the dielectric was characterized by measuring leakage current on a planar film and also by determining the ability to hold voltage in a liquid crystal test cell commonly known as voltage holding ratio or VHR tests. The test configurations are shown in Fig. 2.2. During actual material development, the easiest test to reproduce for the material developer is the leakage current test which employs a typical mercury probe in contact with the dielectric film formed on silicon wafer, creating capacitor. If the mercury-sample contact is ohmic (non-rectifying) then current–voltage instrumentation can be used to measure leakage currents, or current–voltage characteristics. The same mercury-sample structures can be measured with capacitance–voltage instrumentation to monitor permittivity (dielectric constant) and thickness of dielectric materials so the measurements are a convenient gauge for development of novel dielectrics. The VHR test is the most important test to evaluate the dielectric encapsulant. The test is done on a fabricated liquid crystal test cell with stacks of liquid crystal, dielectric and the metal to form contacts. A fixed bias is applied between two metal electrodes encompassing liquid crystal and dielectric and the voltage monitored with time. A good dielectric does not affect the voltage held, so that there is no degradation in the polarization (and therefore display image quality) of the liquid crystal.

Both tests suggest that either conductivity issues or dielectric constant issues are involved in signal loss, but the tests themselves do not help the material developer understand root issues of performance. Specifically for the performance

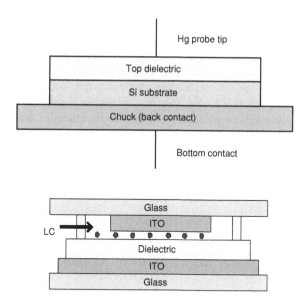

**Fig. 2.2** *Top* Test configuration for leakage current; *Bottom* Test cell for voltage holding ratio (VHR)

of the dielectric used next to the liquid crystal layer (LC), a distinct concern is identifying the major mechanism responsible for any signal loss in order to develop dielectrics that do not impact the performance of the display LC. So although tests are well-known, the ties to the material structures which would help guide material development are absent and molecular modeling was engaged to help fill the gaps in understanding.

Many examples exist in the literature using quantum mechanics for screening conducting materials [2, 3], but fewer references are associated with issues surrounding the electrical performance of dielectrics in applications. Molecular modeling is not usually used for investigating conduction problems in dielectrics. For leakage current it was decided that value might be placed on looking at relative band diagrams (relative conductivity potential) to look for trends in band gap and mobility. As will be discussed in this paper, it was found that exact band structures are not necessary to understand what is happening to the material, as long as adequate benchmarks of known materials have been calculated and comparisons are kept within groups. In this manner, the calculations of interest may be simple, without requiring large unit cells or high levels of theory and we have found that building simple band structure trends are useful in understanding the current leakage trends in our materials under development. For VHR, it was also found that molecule polarization was of great help to correlate the testing to that of the material, as it was decided that this quality best represented the molecular structural quality to respond to an external field.

In all cases, CASTEP, a plane-wave pseudo-potential code based on density functional theory (DFT) from Accelrys Inc. [4–9] has been used employing ultrasoft norm-conserving pseudo-potentials. The generalized gradient approximation (GGA) exchange-correlation was used along with the PBE local density functional, Gaussian smearing and Pulay density-mixing. All calculations made use of periodic unit cells, in which the bonding was adjusted to maintain infinite bonding in at least two, if not all three dimensions.

It should be emphasized that since the modeling was not meant to determine exact values all comparisons made were for directional purposes only. Care has been taken to compare comparable structures to minimize variation in structure outside of the subject query. In addition because of the proprietary nature of all formulations specific structures concerning the dielectric materials under development cannot be reported.

## 2.2 Results

### 2.2.1 Model Benchmarks to Experiment

In order to identify trends in calculations, simple experimental cases were used to benchmark the leakage current trends by looking specifically for bandgap changes in the band structure. In addition, cases from the literature were used to benchmark calculated molecular polarizability (for VHR).

One well-known benchmark that was qualified was the effect of silanol content. It is known that the presence of SiOH (silanol) groups increase the leakage of silicate films; usually found when films are not cured or cross-linked completely as shown in Fig. 2.3. As shown in Table 2.1, the effect of low-temperature curing which enhances silanol content is to decrease the breakdown voltage (at 1 $\mu$A and 1 $\mu$A/cm$^2$ in the last two columns) represented as field to breakdown (FBD) expressed in MV/cm and as a result the films have increased leakage current.

Example structures were built for band-structure calculations, which involved full silicate bonding and then bond cleaved to form partial silicate bonding, terminated in SiOH groups (Fig. 2.4). As expected, qualitatively there is a difference between the band structures, showing that when SiOH is present, the bandgap shrinks (as defined by the gap between the valence band at zero, and the conduction band, which is the next band above it) from around 6 eV to around 4 eV. As further indication that conditions encouraging leakage are being formed, the valence band curvature is increasing suggesting an increased hole mobility.

In addition, another comparison was made using well-known substitution effects observed in-house. Table 2.2 shows the test comparisons of two different functional groups that have been commonly used during dielectric development. Testing in our laboratories found that group R1 will consistently leak and R2 will not. Figure 2.5 shows the calculated bandstructures starting with the same structure using R1 (*left*), then substituting R2 (*right*) for R1. In this comparison, the bandgaps seem to be similar. Because of the size differences in the R1 and R2 groups, the bandgap comparison was further qualified by adjusting densities using

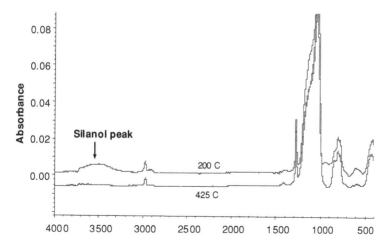

**Fig. 2.3** FTIR of differently cured SOG films showing the higher SiOH region for the lower-temperature cures

**Table 2.1** Effect of cure on field breakdown

| # Cure temp | Dielectric constant, k (Hg probe) | FBD measured at constant current 1 A (MV/cm) | FBD measured at constant current density 1 A/cm² (MV/cm) |
|---|---|---|---|
| 1  250°C/60 min | 3.44 | 3.28 | 3.04 |
| 2  250°C/60 min | 3.43 | 3.34 | 3.08 |
| 3  425°C/60 min | 3.11 | 3.78 | 3.57 |
| 4  425°C/60 min | 3.11 | 3.64 | 3.51 |

different configurations (cage and opened-cage rings) so that the most stable unit cell (lowest energy) was obtained for both R1 and R2. When these band structures were compared, the conclusion was the same: the bandgaps are the same at around 4 eV. The biggest difference between the structures is that the curvature for the conduction band of R1 is highly accentuated suggesting a much higher electron mobility.

These two examples suggest that as we look at comparisons, both the band gaps and the mobility (as inferred by the curvature) should be monitored. Because mobility or effective mass became a metric to monitor, benchmarks for mobility was also estimated based upon curvature slope in the band structures (estimation technique was developed by Accelrys [3–9]). The benchmarks for the mobility trends are found in Table 2.3, showing that qualitatively the direction of mobility is the same as reported in the literature.

The effective masses for all structures obtained for comparing silanol effects, R1 and R2 were calculated and compared to look for general trends; these results are shown in Table 2.3 and Fig. 2.6. As a group, clearly the lowest effective

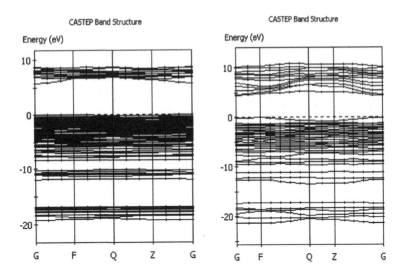

**Fig. 2.4** Band structure of a silicate before (*left*) and after (*right*) bond cleavage to form SiOH groups

**Table 2.2** Experimental tests of group R1 versus R2

| # Refractive index | Dielectric constant, k | FBD measured at constant current 1 μA (MV/cm) | FBD measured at constant current density 1 μA/cm² (MV/cm) |
|---|---|---|---|
| 1 250°C/60 min | 3.44 | 3.94 | 2.62 |
| 2 250°C/60 min | 3.43 | 3.61 | 3.27 |
| 3 425°C/60 min | 3.11 | 4.47 | 4.41 |
| 4 425°C/60 min | 3.11 | 4.47 | 4.42 |

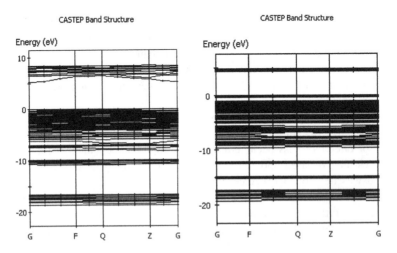

**Fig. 2.5** Band structures of silicates with R1 (*left*) versus R2 (*right*)

**Table 2.3** Effective mass estimates compared to the literature

| | Literature [10, 11] | | Calculated | |
| --- | --- | --- | --- | --- |
| | Effective mass (me) | | Effective mass (me) | |
| | Hole | Electron | Hole | Electron |
| Si | 1.56 [10] | 1.08 [10] | 0.74 | 1.76 |
| Ge | 0.37 [10] | 0.56 [10] | 0.65 | 1.13 |
| $In_2O_3$ | 0.6 [11] | 0.3 [11] | 0.72 | 0.22 |

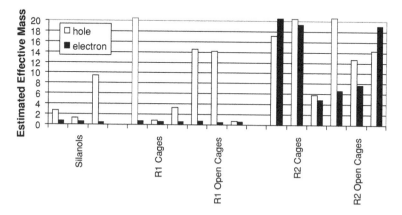

**Fig. 2.6** Effective masses compared across functional group types

**Table 2.4** Polarizability of PVDF

| | CHEMOS: Hartree–Fock, CNDO [12] | CASTEP: DFT GGA/PBE | |
| --- | --- | --- | --- |
| | $(A^3)$ | Optical $(A^3)$ | Static $(A^3)$ |
| PVDF (all trans) | ~9.05 | 9.03 | 11.1 |
| PVDF (cis/trans) | ~8.8 | 8.07 | 10.3 |

masses and so the most conductive are those grouped with silanol content, followed by the R1 structures. The R2 structures, as a group definitely have the highest effective masses and so overall should have the least issues with leakage current.

As will be discussed later, polarizability calculations were found to be specifically important to understand VHR and the calculations were first benchmarked against the literature using a well-known piezoelectric polymer, polyvinylidene-fluoride (PVDF). As shown in Table 2.4, even though different methods are used from the literature, the current calculations adequately represent the decrease in polarizability as the cis isomer is introduced and there is high agreement in values for the trans isomeric form with exact matches in structure.

**Table 2.5** Leakage results for two different Formulations A and B

| Formulation | Refractive index | Dielectric constant, k | FBD measured at constant current 1 μA (MV/cm) | FBD measured at constant current density 1 μA/cm² (MV/cm) |
|---|---|---|---|---|
| A | 1.5002 | 3.48 | 4.58 | 2.09 |
| A | 1.5009 | 3.50 | 4.32 | 2.48 |
| B | 1.5115 | 3.83 | 3.69 | 1.76 |
| B | 1.5114 | 3.84 | 3.59 | 1.75 |

## 2.2.2 Leakage Comparisons to Bandstructure

The most important aspect of this work was to determine the electrical performance trends of actual formulations, and there were two major formulation types that were of concern. The leakage results are shown in Table 2.5, obviously showing that Formulation B has much lower breakdown voltage, and so should be highly leaky.

The band structures of these two formulations were obtained, using different densities to ensure that a good range of possibilities were considered. Because of a basic functional group difference in one of the components in Formulation B, densities were obtained assuming a low-cure (cure 1) and a high-cure (cure 2) situation with different extents of cross-linking. The results are compiled in Fig. 2.7. As a group, formulation B consistently has lower bandgap properties than formulation A. It appears that if formulation B attains a higher cure, and higher densities, the band gap may suddenly drop. The effective masses were determined and represented in Table 2.6. In general the effective masses are high for both formulations, except for one case in Formulation B where the effective masses are low enough to have serious concerns for leakage. As mentioned previously, leakage current experiments suggest that Formulation B is highly leaky. The decision based on this result would have been not to use this material in any device that needs this functional requirement.

To further test differences between the two formulations and to see whether there are any unexpected interface quality issues with the liquid crystal (LC) or polyimide (PI) interface that might be present in the actual display, the interaction energies of each formulation were calculated with test cases derived from the literature that might be representative of possible polymers (LC or PI) used in device fabrication to test whether there is a fundamental difference between interfaces made with Formulation A and B. Two test cases were used for each polymer shown in Fig. 2.8, where unit cells were built from a layer of the polymer and a layer of the formulation. The formation energies of this interface were compared to formation of a free surface (Fig. 2.10). In all cases, Formulation B appears to have a weaker interface. To further check on the electrical properties of the interface, the band gaps and effective masses of the interface unit cells were determined. These results are found in Table 2.7, showing that in general the band

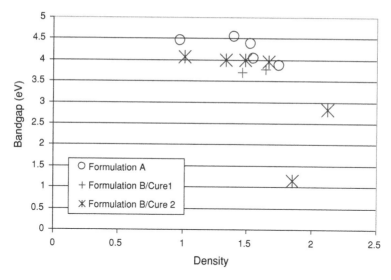

**Fig. 2.7** Band gap (Fermi to conduction band gap in eV) comparisons of Formulations A and B

**Table 2.6** Effective masses (mobility) for Formulations A and B

|                     | Density | Bandgap (eV) | Hole effective mass | Electron effective mass |
|---------------------|---------|--------------|---------------------|-------------------------|
| Formulation A       | 1.5     | 4.38         | 84.01               | 21.34                   |
| Formulation A       | 1.6     | 4.03         | 27.65               | 2.82                    |
| Formulation B/cure1 | 1.6     | 3.78         | 152.45              | 5.87                    |
| Formulation B/cure1 | 1.5     | 3.71         | 65.59               | 21.21                   |
| Formulation B/cure2 | 1.7     | 3.97         | 11.03               | 14.61                   |
| Formulation B/cure2 | 1.4     | 3.99         | –                   | 0.88                    |

gaps are similar (it may be argued that Formulation B has a slightly lower band gap) and all interfaces have relatively high-effective masses, so do not have mobile carriers. That is, there is no reason to believe that the interface contributes to the leakage due to carriers.

## 2.2.3 VHR and Polarization Comparisons

The most important electrical test for acceptance of the dielectric encapsulant is the voltage holding ratio (VHR) test. As explained previously it requires the fabrication facility to prepare a test cell and is more time consuming. The results for formulation A and B are found in Fig. 2.9. These figures show that films from Formulation B are better at holding the voltage bias, which is surprising since Formulation B showed higher leakage tendencies.

**Fig. 2.8 a** Two common LC structures used to test formulation interface stability. (*Top* is designated LC1, and *bottom* is designated LC3). **b** Two common polymimide structures used to test formulation interface stability (*Top* is designated PI1, and *bottom* is designated PI5)

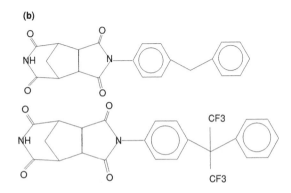

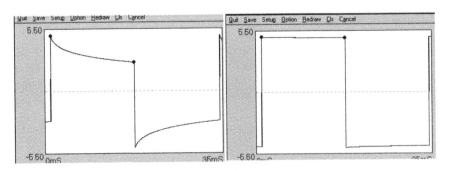

**Fig. 2.9** VHR test results. *Left*: Formulation A; *Right*: Formulation B

Since voltage holding is directly related to how well a dielectric may hold its charge, which is in turn related to how well a structure may hold polarization, the polarizablities of the formulations were determined. Figure 2.10 shows the calculated polarizabilities of each formulation using several different densities; for Formulation B the high- and low-cure structures were also used. Consistent with

**Table 2.7**  Relative effective masses of interface structures

| Interface representation | Bandgap | Effective mass estimate (me) | |
|---|---|---|---|
| | | Hole | Electron |
| Formulation A surface | 4.38 | 39.25 | 54.64 |
| Formulation A-LC1 | 3.95 | 6.69 | 19.38 |
| Formulation A-LC3 | 3.47 | 87.60 | 237.12 |
| Formulation A-PI1 | 3.31 | 317.72 | 10.74 |
| Formulation A-PI5 | 3.6 | 41.19 | 70.71 |
| Formulation B/cure2 surface | 3.93 | 5.03 | 11.27 |
| Formulation B/cure2-LC1 | 3.3 | 28.73 | 18.62 |
| Formulation B/cure2-LC3 | 3.31 | 23.45 | 29.88 |
| Formulation B/cure2-PI1 | 3.19 | 87.12 | 21.51 |
| Formulation B/cure2-PI5 | 3.22 | 17.67 | 262.14 |

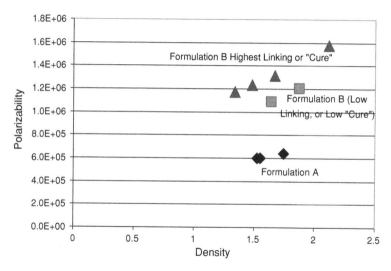

**Fig. 2.10**  Polarizability of Formulations A and B (B is shown for two different cure states)

the stable VHR, Formulation B consistently has the higher polarizability. This trend also agrees with the higher measured dielectric constant of Formulation B than Formulation A (Fig. 2.10).

## 2.3  Discussion and Conclusions

When compared against the experimental results, the modeling results appear to be consistent within tests. That is, leakage trends are consistent with bandgap trends and estimated effective masses hint at which structures may be more susceptible to leakage; and the voltage holding ratio tests are consistent with the polarizabilities of the structures, showing the importance of polarity within the structures adding to the overall capacitance.

However, the two results seem counter-intuitive to one another if it assumed that leakage is a major detractor to switching performance and display instability while high VHR indicates better performance. These results suggest that for these dielectrics the more important contribution comes from the native capability for polarization and so it performs as a better dielectric in contact with liquid crystal. In addition, charge carrying capability does not appear to be significant for the dielectric's role judging from similarly high-effective masses for both formulations. So, the higher polarizability of Formulation B and so its higher dielectric constant suggests that it should perform like a better dielectric and higher voltages should be sustained before carriers can be developed, dielectric breakdown occurs and conduction mechanisms are dominant. However the recent models and tests do not answer when such polarization/capacitance mechanisms give rise to internal carrier generation and leakage, or at what point this mechanistic change becomes important to the display performance. Interestingly, Formulation B also has less interface interaction between the LC and the dielectric. Looked at in another way, the dielectric might help LC phase transformation just because of lower friction/stiction. If this is the case there are clearly more mechanistic issues that could be addressed from the film interaction aspect [13, 14] including, the role of adhesion and wetting at the interface and what balance must exist at the layer interfaces for the dielectric to help LC performance without layer delamination.

Overall, the modeling shows that it can be used to help separate performance effects from the structural level, as well as help understand issues from the device side. For instance these calculations did help to highlight the polarization effects and show that these effects may be related to VHR, further separating the mechanism from those effects derived from conductivity to carrier generation. Interestingly the leakage tendencies do match bandgap and mobility trends, even though such calculations are reserved for more conductive materials, so as a qualitative tool for non-conductive materials, the calculation can still be used. Generally, we have found that such models used to gauge electrical effects do not have to be quantitatively perfect in order to aid the material developer, but care must be observed in comparisons (benchmarks used) so that the emerging trends may be trusted.

**Acknowledgments** The authors would like to acknowledge discussions held with Stephen Yates and Kenneth Heffner of Honeywell Aerospace that helped us justify looking into at band-structure trends for leakage susceptibility. Although all statements and information contained herein are believed to be accurate and reliable, they are presented without guarantee or warranty of any kind, express or implied. Information provided herein does not relieve the user from the responsibility of carrying out his own tests and experiments, and the user assumes all risks and liability for use of the information and results obtained. Statements or suggestions concerning the use of materials and processes are made without representation or warranty that any such use is free of patent infringement and are not recommendations to infringe any patent. The user should not assume that all toxicity data and safety measures are indicated herein or that other measures may not be required.

# References

1. Lee JH, David NL, Shin-Tson Wu (2008) Introduction to flat panel displays. Wiley, UK
2. Ren CY, Chiou SH, Choisnet J (2006) First-principles calculations of the electronic band structure of $In_4Sn_3O_{12}$ and $In_5SnSbO_{12}$. J Appl Phys 99:023706
3. Zgou H, Bouzzine SM, Bouzarkraaoui S, Hamidi M, Bouacrine M (2009) Theoretical study of structural and electronic properties of oligo(thiophene-phenylene)s in comparison with oligothiophenes and oligophenylenes. Chin Chem Lett 19:123–126
4. Accerlys, Inc. San Diego, CA (CASTEP was employed under Materials Studio 4.3 for these calculations)
5. Segall MD, Lindan PJD, Probert MJ, Pickard CJ, Hasnip PJ, Clark SJ, Payne MC (2002) First-principles simulation: ideas, illustrations and the CASTEP code. J Phys: Cond Matt 14:2717–2744
6. Kohn W, Sham LJ (1965) Self-consistent equations including exchange and correlation effects. Phys Rev 140:A1133
7. Vanderbilt D (1990) Soft self-consistent pseudopotentials in a generalized eigenvalue formalism. Phys Rev B 41:7892–7895
8. Perdew JP, Burke K, Ernzerhof M (1996) Generalized gradient approximation made simple. Phys Rev Lett 77:3865–3868
9. Alibert C, Skouri M, Joullie A, Benouna B, Sadig S (1991) Refractive indices of AlSb and GaSb-lattice matched AlxGa1-xAsySb1-y in the transparent wavelength region. J Appl Phys 69:3208–3211
10. Sze Simon M, Kwok K Ng (2001) Physics of semiconductor devices. Wiley, Chichester
11. Zheng MJ, Zhang LD, Zhang XY, Zhang J, Li GH (2001) Fabrication and optical absorption of ordered nanowire arrays embedded in anodic alumina membranes. Chem Phys Lett 334:298–302
12. Correia Helena MG, Ramos Marta MD (2005) Comput Mater Sci 33:224–229
13. Iwamoto N, Krishnamoorthy A, Spear R (2009) Performance properties in thick film silicate dielectric layers using molecular modeling. Microelectron Reliab 249:877–882
14. Iwamoto N, Li T, Sepa J, Krishnamoorthy A (2007) Dielectric constant trends in silicate spin-on-glass. SPIE Nanoengineering: fabrication, properties, optics and devices IV session, conference 6645, San Diego, 26–30 August 2007

# Chapter 3
# Understanding Cleaner Efficiency for BARC ("Bottom Anti-Reflective Coating") After Plasma Etch in Dual Damascene Structures Through the Practical Use of Molecular Modeling Trends

**Nancy Iwamoto, Deborah Yellowaga, Amy Larson, Ben Palmer and Teri Baldwin-Hendricks**

This paper is based upon "Using Molecular Modling to Understand Cleaner Efficiency for BARC ("Bottom Anti-Reflective Coating") Layers after Plasma Etch in Dual Damascene Structures" which appeared in the Proceedings of Eurosime © 2007, IEEE.

**Abstract** This work describes the surface energy and reactivity modeling of the Honeywell cleaner technology that has demonstrated an inorganic Bottom Anti-Reflective Coating removal solution with significantly higher removal rates and etching selectivity compared to currently available products. We will show how molecular modeling helped to influence the development of the cleaner formulation and the impact on experiments.

## 3.1 Introduction and Background

The invention of inorganic bottom anti-reflective coatings (iBARC) was a promising enabling technology for lithography due to their near-unity plasma etch selectivity with the low-k dielectric materials used in advanced via first trench last (VFTL) dual damascene structures. Having these selectivities during plasma etch allowed for the avoidance of via "fence" defects that are often created by the shadowing of the dielectric surrounding the vias by the slower etching organic BARC (oBARC) via fill. The presence of "fence" defects creates an area of high

N. Iwamoto (✉) · D. Yellowaga · A. Larson ·
B. Palmer · T. Baldwin-Hendricks
Honeywell Electronic Chemicals, 6760 W. Chicago St,
Chandler, AZ 85226, USA
e-mail: nancy.iwamoto@honeywell.com

N. Iwamoto et al. (eds.), *Molecular Modeling and Multiscaling Issues for Electronic Material Applications*, DOI: 10.1007/978-1-4614-1728-6_3,
© Springer Science+Business Media, LLC 2012

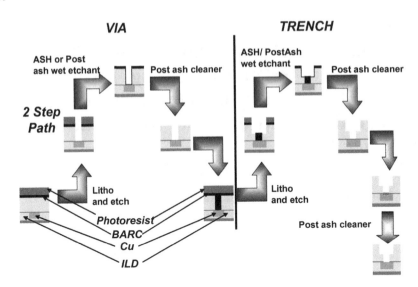

**Fig. 3.1** Process flow for inorganic BARCs (iBARC) for dual damascene structures

stress in the copper line due to the tight space between the fence and side of the trench that has to be filled, and can cause failure of the device due to shorting of the line from voids caused by copper migration away from the high stress area. In order to achieve a 1:1 selectivity in plasma etch which avoids fence formation, organosiloxane polymers are typically employed in the inorganic "Bottom Anti-Reflective Coating" (BARC) materials. However, modification of the organosiloxane film during plasma etching and ashing densifies the material and may also remove organic content from the film, making removal more difficult with traditional cleaning technologies.

"Bottom Anti-Reflective Coatings" are used as lithography enablers during dual damascene structure build up. The BARC is added under the photoresist to enhance exposure, and is removed during development and etch/clean. For the BARC to perform properly it must have differential reactivity in order to distinguish it from the interlayer dielectric so that clean and sharp features result from the process [1]. The general steps used for inorganic BARCs is found in Fig. 3.1 showing the location of the BARC, photoresist and interlayer dielectric (ILD). As may be appreciated, the BARC should not only develop or etch worse than the ILD to avoid undercutting, but also must be completely removed in order to open features as seen on in the upper processes by ashing or through the use of newly developed Honeywell post-ash etchants and cleaners in Fig. 3.1.

One issue that has emerged is the tendency of iBARC layers to densify after the $O_2$ plasma etch step which affects the BARC's reactivity toward further removal by wet etchants and cleaners. Another experimental issue that has emerged is the observation that upon aging, the iBARC layers become more resistant to wet etchants and cleaners. This work will show that the iBARC (iBARC 193, and iBARC 248) developed by Honeywell as well as the Honeywell etchants and cleaners have been tuned together to maximize removal after the $O_2$ plasma, and how the structural nature of the iBARC impacts its removal ability and explains both the removal of the $O_2$ plasma densified films and the aging data.

In order to gain the fundamental understanding of the trade offs present between the iBARC and the etchant, molecular modeling was applied in several scales. The smaller, atomistic scale looked at relative reactivities between the major molecular frameworks thought to be present from analysis of the films. A higher scale molecular surface model investigated the relative wetting of the etchant with the target film; and a combined technique used the charge balances from the atomistic models to tune surface simulations for wetting tendencies. This work will also demonstrate how relative thermodynamic models (as opposed to quantitative models) can be used constructively within a development mode.

## 3.2 Investigations of BARC Etching Tendencies

### 3.2.1 Thermodynamic Model Results

As mentioned previously, after $O_2$ plasma etch, the iBARC tends to densify and change its' reactivity. In order to understand the change in reactivity, a combination of thermodynamic models (using the DFT/quantum mechanics code DMOL [2]) in conjunction with FTIR analysis was run to understand the underlying changes, and the reasons behind the performance observations. In all cases free energies of reaction were calculated employing the DNP basis set and the GGA/BP functionals.

For the reactivity modeling, general structures that represent network formation were used to understand reactivity of the iBARC, by specifically calculating the formation thermodynamics of network elements, and both SiOSi bond formation and hydrolysis trends. These elements are represented in Fig. 3.2. mostly representing cage (middle four structures) and ladder (last two right structures). Most of the work was geared toward the investigation of the reactivity of the cages and ladders, as representatives of network structures. By doing so four, five and six membered ring structures are represented. For the hydrolysis steps (for investigation of etch and cleaning) and condensation steps (for investigation of effects from the $O_2$ plasma as well as aging), usually the last step of the structure formation is calculated to give an indication of tendency of reaction of the different network components.

Figures 3.3 and 3.4 summarize the hydrolysis tendencies of the structures as identified by the calculated free energies of reaction. Figure 3.3 investigated the tendency of hydrolysis due to organic content, as represented by ring cleavage of the small four silicon ring unit. It was found that in general, the higher the organic content of the iBARC and the lower the silanol content, the structures are more resistant to hydrolysis as indicated by an increase in the expected endothermicity of the reaction with higher organic content. For the larger structures, it was found that the larger cages and ladders with larger unstrained multicyclic rings, such as

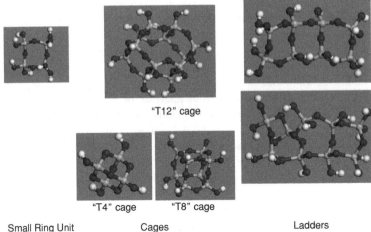

"T12" cage

"T4" cage        "T8" cage

Small Ring Unit          Cages                    Ladders
                                        (lower stress ladder at bottom)
**Fig. 3.2** General structural units investigated

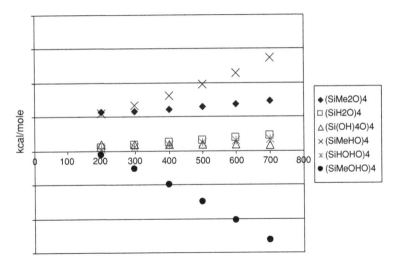

**Fig. 3.3** Relative free energies demonstrating the effect of organic content in iBARCS on hydrolysis of ring opening (relative free energies)

the T8, T12 structure (middle top cage structures in Fig. 3.2, T8 on left and T12 on right) and large ladder (bottom ladder structure in Fig. 3.2) are the most resistant to hydrolysis. For condensation that might occur during aging or plasma etch, the same hydrolysis models were used by analyzing for the inverse reaction. In doing so, it was concluded that the higher the organic content the higher the tendency to condense to network structures; and the more strained the resulting network structure, the lower the tendency to form. However, the small T4 branched (lower left cage structure in Fig. 3.2) structure has low enough free energy at higher temperatures that it might form at higher temperatures (Fig. 3.4, "T4-O-T4").

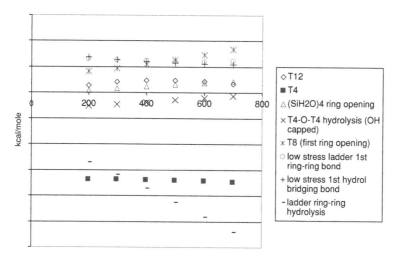

**Fig. 3.4** Relative free energies demonstrating the effect of structure in iBARCS on SiOSi hydrolysis

These general trends were then compared with FTIR analysis, which was applied to look for the structural changes which will be discussed in the next section along with the experimental studies. Together with the thermodynamic tendencies, the structural changes make clear the reasons behind the experimental observations of etch behavior.

## 3.2.2 Experimental Film Analysis and Correlation to the Modeled Thermodynamics

The thermodynamic modeling of structural comparisons was particularly important in the understanding of several aspects of iBARC experimental film stability, including the effect of $O_2$ plasma etch and the issue of film aging. For example, for $O_2$ etched films, the FTIR analysis of iBARC 193 and iBARC 248 films have shown that both 193 and 248 forms of the BARC react to a similar structure, as demonstrated in Fig. 3.5. (before treatment) versus Fig. 3.6. (after $O_2$ plasma etch). These FTIR differences indicate that the main effect of the $O_2$ plasma treatment is that organic content was eliminated in both types of iBARC films as shown in the loss of peaks around 1270 and 1130 cm$^{-1}$.

The impact of such structural changes was indicated by the thermodynamic models that suggested that organic loss should lower the resistance to reaction, and so lower the resistance to the wet etch. The observed experimental increase in wet etch (found in Table 3.1) is mainly due to loss of the organic, in agreement with both the FTIRs of the films and the tendencies predicted by modeling.

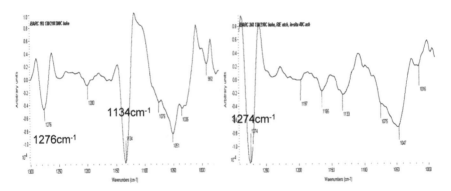

**Fig. 3.5** FTIR absorbance examples of iBARC films before O$_2$ Plasma Etch. An iBARC 193 film is shown at *left* and an iBARC 248 film is shown at *right*

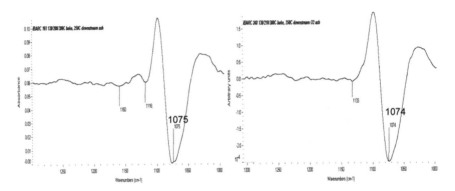

**Fig. 3.6** Post-O$_2$ plasma etch FTIR absorbance example. An iBARC 193 film is shown at left and an iBARC 248 film is shown at right

**Table 3.1** Wet etch performance of post-O$_2$ plasma iBARC 248 films

|  | Etch rate, angstroms/min |
| --- | --- |
| 130/210°C 60 s bake | 2085.2 |
| 130/210°C 60 s bake, 60 s O$_2$ (Gasonic) | 7755 |
| 130/210/250°C 60 s bake, 60 s O$_2$ (Gasonic) | 6810.5 |

However, because plasma etch also involves high local temperatures there may be more different types of network formation, including the higher strained content that may add to the reactivity of the film. The broadening of the SiOSi peak suggests that network formation is occurring which can include such structures.

For aged films, there are subtle differences in the FTIR before and after aging. However there is an increase in the OSiO intenstity at $\sim 1050$ cm$^{-1}$ which suggests

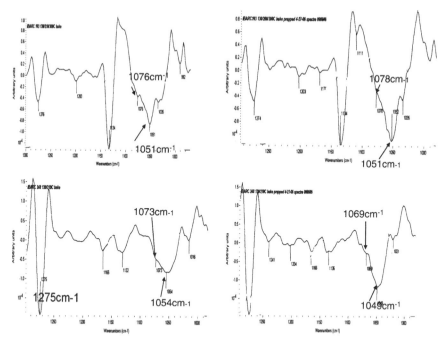

**Fig. 3.7** FTIR absorbance of iBARC before (*left*) and after (*right*) aging. *Top*: iBARC 193; *bottom*: iBARC 248

longer chain and less strained network structures (Fig. 3.7) is present which according to the thermodynamic trends between ring types should enhance the resistance of the film to hydrolysis. There is also a difference between the 193 and 248 iBARC films. That is, the iBARC 193 maintains a high quantity of cage structures (higher absorption at $\sim 1130$ cm$^{-1}$) after aging which again according to the thermodynamic trends, should increase the 193 film resistance to hydrolysis over the iBARC248.

Experimentally, it is found that indeed the iBARC 193 films in general take longer to etch as shown in the top chart of Fig. 3.8, and is in direct agreement with both the calculated thermodynamic trends as backed by the FTIR analysis of the films. It is also found that the aged films are more resistant to etch, and this is shown in the bottom example of Fig. 3.8 for an iBARC 248 film.

## 3.2.3 Investigations of the Etchant/Cleaner

In addition to structural effects of the iBARC, the other material that was investigated was the effect of the Honeywell etchant/cleaner on efficiency. For the modeling work both the wetting ability and the reactivity of the etchant was

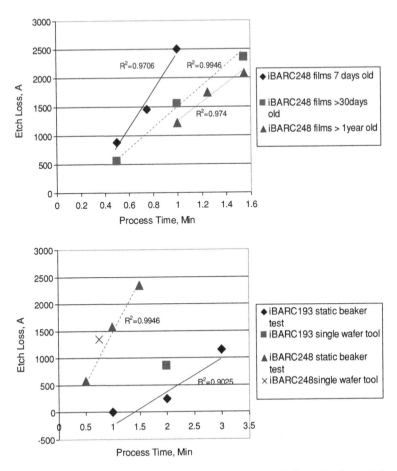

**Fig. 3.8** Wet Etch performance of aged films. (*Top*) Example of effect of aging and (*bottom*) iBARC 248 versus iBARC 193

investigated. While specifics of the formulation cannot be divulged, the basic trends will be reported here.

For the wetting tendencies, the basic Honeywell etchant/cleaner was modeled on several different surfaces using molecular modeling techniques previously applied to adhesives and underfills [3–6] in which the surface energy is qualitatively determined by energy drop trends during room temperature dynamics after liquids are introduced to a surface. (Quantitative surface energy determination makes use of calibration that is outside the scope of the current work.) The wetting simulations used to derive the relative wetting or surface energy trends were obtained using molecular dynamics as applied in Discover/Insight available from Accelrys, in [2]. The etchant/cleaner was contrasted for wetting behavior on a simulated iBARC (Fig. 3.9) and also two different (inner layer dielectric) ILD

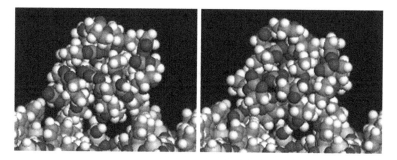

**Fig. 3.9** Before/after wetting simulation of the basic Honeywell etchant/cleaner formulation on iBARC 193 showing a larger area interface after wetting between the etchant/cleaner and the iBARC

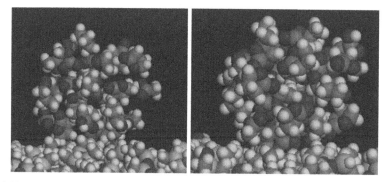

**Fig. 3.10** After wetting on TMCTS low-k (hydride/organic-based), *right*; and after wetting on OMCTS low-k (organic-based), *left*

structures, Fig. 3.10. Before/after images of the wetting models on iBARC are shown in Fig. 3.9 and indicate that there is better wetting on the iBARC because of the that higher intimate contact is achieved between the Honeywell cleaner-simulant and the surface. Because contact is a prerequisite to higher reactivity and less residual formation, this formulation is expected to perform better on the iBARC 193 surface than on other silicates used for ILD layers, and should then give selectivity between surfaces.

The reason for the differentially better wetting of the iBARC is due to a balance of charges during the formulation process of the etchant/cleaner leading to the low surface energies required for wetting (Fig. 3.11 middle bar).That is, either too large of ionization or too little will result in an increase in surface energy for the etchant/cleaner which will hurt wetting. The interesting consequence of this balance of charges is not only an increase in wetting, but also an increase calculated reactivity (Figs. 3.12 and 3.13). Figure 3.12 shows a variety of structures and etchant environments and free energy of hydrolysis. According to the data, in general, increasing the ionic content helps the hydrolysis, but it is not a smooth

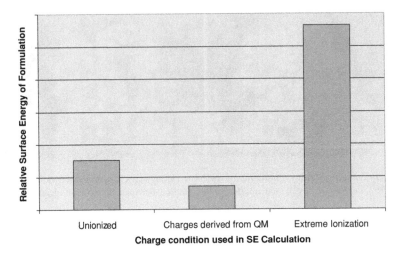

**Fig. 3.11** Effect of charge content on the relative surface energy of the Honeywell etchant/cleaner

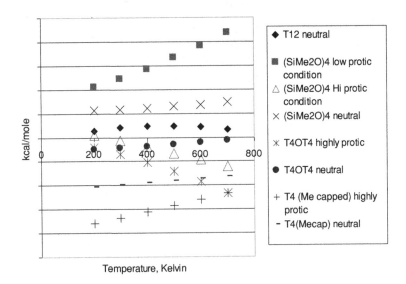

**Fig. 3.12** Relative free energies of ring hydrolysis for general ring types showing how charge structure may improve free energy

relationship. That is, identification of the type of ion does not necessarily mean that the reaction is more exothermic, rather the overall effect of the ionization on the formulation must be considered as in Fig. 3.11. Figure 3.13 shows free energy of hydrolysis of the structural components of a typical ILD (left bars) and the components of the iBARC. Using the Honeywell etchant/cleaner, it appears that the energetic of reaction are tuned more for the iBARC than for the ILD. So, from

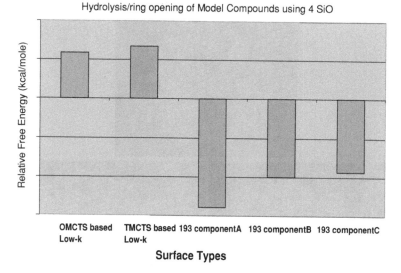

**Fig. 3.13** Relative reactivity differences of ILD materials versus iBARC 193 using a model 4Si ring opening reaction

comparisons of the modeling results of the surface energies and the reactivities, both the surface energy of the etchant/cleaner as well as the reactivity of the formulation have been tuned in order to wet the iBARC preferentially as well as react better to the iBARC.

In addition, the Honeywell formulations were tested experimentally against competitor cleaners. Results were found to exceed the selectivity of other cleaners and in-line with expectations of the specific performance tuning from the modeling [1].

Using simple molecular models to establish basic performance trade offs was found to be convenient to fast-track tuning experiments for new etchant break thoughs. For instance, Figs. 3.14, 3.15, 3.16 provide the results of another example with completely different design goal requirements in which the oxide rates needed to be equivalent regardless of their doped state, but faster than silicon. Figure 3.14 provides some of the basic wetting guidance found in the modeling, showing that thermal oxide (silica) wets significantly differently than BPSG (boron/phosphorous doped silica) or PSG (phosphorous doped silica), and so the formulation had to be tuned differently to take into account the differential wetting tendencies. With the probe formulation model, wetting trends were expected to be silica > BPSG > PSG, but different mixtures were tested that changed the extent of wetting in the models to determine direction. This information was fed into an experimental mixture design optimization plan (Fig. 3.15 shows the results of the mixture design scheme) which quickly located the optimum variable space. Tests of the optimized mixture (Fig. 3.16) showed that the experimental design goals

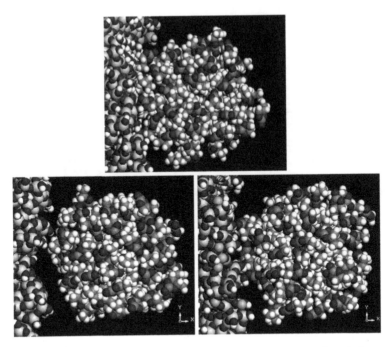

**Fig. 3.14** Wetting simulation of different oxides using a probe etchant formulation to gain understanding of the fundamental oxide differences. *Top*: Thermal Oxide (silica); *bottom left*: BPSG (Boron/Phosphorous doped Silica); *bottom right*: PSG (Phosphorous doped Silica)

**Fig. 3.15** Mixture design. Optimum area is indicated by the *star* showing that a differences between polysilicon and the different oxides, thermal oxide (TOx), BPSG and PSG

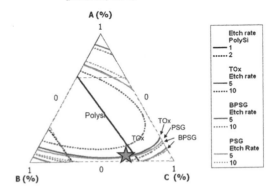

were met. As is demonstrated by Figs. 3.15 and 3.16, this performance target was distinctly different from the previous example which had design goals of extreme selectivity between two iBARC materials (Fig. 3.8).

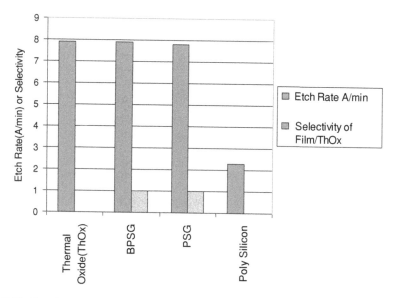

**Fig. 3.16** Experimental results of optimization showing maximum selectivity between the Oxides, Thermal Oxide (ThOx), BPSG, BPS and Polysilicon

## 3.3 Conclusions

The modeling work described in this paper demonstrates how modeling was used as an aid in understanding the experimental work in developing a new etchant/cleaner for iBARC layers. The modeling demonstrated how the structural change to both the iBARC and the etchant formulation impacts the performance of the film.

It was found that aged iBARC films are expected to have higher resistance to the wet etch because of the formation of lower strained networks/rings and that differences in etching between types of iBARCs upon aging are predictable based upon the types of ring structure, with the iBARC 193 more resistant to etchants than iBARC 248 based upon its' higher cage composition. In addition it was found that higher wet etch ability after plasma etch can be attributed directly to lower organic content. By modeling the etchant itself, it was shown that the Honeywell etchant has been tuned to maximize its' wetting properties to the surface of the iBARC, as well as maximizing its' chemistry to react preferentially to the iBARC rather than the ILD.

In order to gain a complete picture, a combination of quantum mechanical (thermodynamic) models which rely on electron distribution as underlying principles as well as classical mechanics (for the surface energy modeling) which rely on classical force field definition were used demonstrating that molecular methods spanning multiple scales are a better way to understand the performance properties.

# References

1. Yellowaga D, Larson A, Palmer B, Baldwin-Hendricks T, Iwamoto N (2006) Highly selective removal of plasma etched and ashed inorganic BARC materials; center for microcontamination control. In: 4th international surface cleaning workshop, Boston, MA, 8 Nov 2006
2. Accelyrs Inc, San Diego CA For the quantum mechanics work the DNP basis set employing the GGA/BP functionals were used
3. Iwamoto N, Li M, McCaffrey SJ, Nakagawa M, Mustoe G (1998) Molecular dynamics and discrete element modeling studies of underfill. Int J Microcircuits Electron Packag 21(4): 322–328 (Fourth Quarter)
4. Iwamoto N, Nakagawa M, Mustoe GGW (1999) Simulation underfill flow for microelectronics packaging. In: Proceedings of the 49th electronic components and technology conference, San Diego, CA, pp 294–301
5. Iwamoto NE, Nakagawa M, Mustoe G (1968) Predicting material trends using discrete newtonian modeling techniques. In: Proceedings of Nepcon West'99, vol III. pp 1689–1698
6. Iwamoto NE, Nakagawa M (2000) Molecular modeling and discrete element modeling applied to the microelectronics packaging industry. In: Micro materials conference, Apr 2000, Berlin, Germany, pp 17–19

# Part II
# Large-Scale Atomistic Methods and Scaling Methods to Understand Mechanical Failure in Metals

## Introduction

This section will show examples of large-scale atomistic models in which the author has begun modeling by enlarging the number of atoms to explicitly model the specific failure sought. Chapter 4 (Shimokawa) shows a good example of such scaling using millions of atoms, for metal grain boundary failure. However Shimokawa goes further in describing quasi-continuum methods to further scale the models. The multiscale aspect of this paper is significant and could also be placed in section V, but because of the dual significance of the molecular modeling aspect it is placed early in this book. Chapter 5 (Sau et al.) deals with specifically understanding crack growth propagation for ultra-fine pitch copper interconnects. Significantly, Sau steps through the aspects needed to simulate the microstructure, the modulus and crack growth mechanisms in multicrystalline copper. Both papers are excellent examples of large-scale atomistic development and its use to look at grain structure.

## Part II Chapter List

**Chapter 4:** "Roles of Grain Boundaries in the Strength of Metals by Using Atomic Simulations"
Tomotsugu Shimokawa

**Chapter 5:** "Semi Emprical Low Cycle Fatigue Crack Growth Analysis of Nanostructure Chip-To-Package Copper Interconnect Using Molecular Simulation"
S. Koh, A. Saxena, W. D. Van Driel, G. Q. (Kouchi) Zhang and R. Tummala

# Chapter 4
# Roles of Grain Boundaries in the Strength of Metals by Using Atomic Simulations

Tomotsugu Shimokawa

This paper is based upon "Atomistic Simulations of Interface Properties in Metals", Tomotsugu Shimokawa which appeared in Proceedings of Eurosime © 2007 IEEE.

**Abstract** Roles of grain boundaries in the strength of ultrafine-grained metals are investigated using conventional molecular dynamics and quasicontinuum simulations. Two problems are considered: (a) grain-size dependence of grain-boundary roles in polycrystalline metals using conventional molecular dynamics simulations and (b) influence of grain-boundary structures on dislocation transmissions through grain boundaries using quasicontinuum method that is a concurrent multi-scale method that couples atomistic and continuum descriptions. A transition from grain-size hardening to grain-size softening can be observed as grain size decreases to nanometer order, i.e., the primary role of grain-boundary changes in each region. It is also found that grain-boundary structures strongly influence the accommodation process for incoming dislocations in the grain boundary and also affect the critical force on the dislocation to eject from the boundary.

## 4.1 Background

In conducting materials, we usually require not only high electrical conductivities but also high mechanical properties, e.g., high yield stress. In order to improve the strength of conducting materials maintaining their purity, grain refinement is a

T. Shimokawa (✉)
School of Mechanical Engineering, College of Science and Engineering,
Kanazawa University, Kakuma-machi, Kanazawa, Ishikawa 920-1192, Japan
e-mail: simokawa@t.kanazawa-u.ac.jp

N. Iwamoto et al. (eds.), *Molecular Modeling and Multiscaling Issues for Electronic Material Applications*, DOI: 10.1007/978-1-4614-1728-6_4,
© Springer Science+Business Media, LLC 2012

very useful strengthening method because ultrafine-grained (UFG) metals with a grain size of the order of 100 nm can now be produced in bulk by severe plastic deformation [1–3]. However, in general, electrical conductivities and mechanical strength are both strongly affected by their internal structures; the trade off for the increase of the volume fraction of grain boundaries, which play a role of obstacles to dislocation movement, is some loss of electrical conductivities. Recently, it has been reported that pure copper samples with the average grain size of about 400 nm and a high density of nanoscale twins show high strength and electrical conductivity, and furthermore show considerable larger tensile ductility than nanocrystalline Cu specimens with few twins [4]. The result gives us motivations to investigate the grain-boundary structure dependence of grain-boundary roles in mechanical properties considering atomic-scale resolutions in order to design conducting materials with high strength and electrical conductivity.

Atomic simulation is a powerful tool for examining grain-boundary properties in detail because this simulation can directly depict the defect structures at the atomic level within the limitations of treatable time and space [5–9]. Furthermore, to overcome the limitations of atomic simulations, a large number of multi-scale methods have been developed [10–12]. In this study, we investigate the roles of grain boundaries in the strength of ultrafine-grained metals by using atomic simulations. Two defect interaction problems are investigated: (1) grain-size dependence of grain-boundary roles in polycrystalline metals by using conventional molecular dynamics simulations and (2) influence of grain-boundary structures on dislocation transmissions through grain boundaries by using the quasicontinuum method [13–15], which is a concurrent multi-scale method that couples atomistic and continuum descriptions.

## 4.2 Grain-Size Dependence of the Grain-Boundary Role in Polycrystalline Metals

### 4.2.1 Models and Methodologies

We consider quasi two-dimensional (2D) nanocrystalline aluminum models with grain sizes $d$ ranging from 5 to 80 nm (Fig. 4.1a, b). The embedded atom method proposed by Mishin et al. [16] is adopted for atomic interactions. This atomic potential can reproduce stable and unstable stacking fault energies of aluminum with accuracy. The analysis models are composed of unit structures that consist of eight hexagonal grains, A–H, in which the crystal orientation along the $X$ direction is fixed at (110), which is the same direction used in the previous quasi 2D models [17]. The length of all the models along the $X$ direction is approximately 1.1 nm, and each model comprises eight (110) atomic planes in the $X$ direction. A periodic boundary condition is adopted for all directions. Due to the geometric restrictions on the simulation cell, 60° dislocations on two slip planes such as (11-1) and (111) as shown

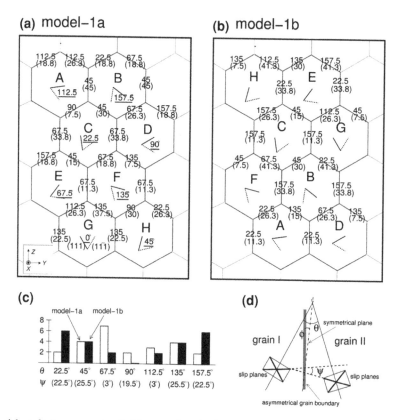

**Fig. 4.1 a, b** Arrangements of eight crystal grains in the nanocrystalline models: model 1a and model 1b. **c** Histogram of the grain-boundary misorientation angle $\theta$ and the minimum angle $\phi$ between the two slip planes in neighboring grains in each model. **d** Geometries of the grain boundary, $\theta$ and $\phi$, and slip systems, $\psi$

by the solid and broken lines in Fig. 4.1a and b fundamentally dominate intragranular deformation. The distance between the two partial dislocations of a 60° dislocation is influenced by the crystal orientation because the resolved shear stresses applied to the leading- and trailing-partial dislocations are not identical [17].

Two types of grain arrangements are considered in order to produce different distributions of the grain-boundary misorientation angle $\theta$; we term these two arrangements as model 1a and model 1b. Both the models have an identical texture because they consist of the same eight grains A–H, hence, it can be easy to investigate the roles of grain boundaries in the strength of polycrystalline metals by comparing results of these models. Figure 4.1c shows the histogram of $\theta$ in both the models, and each value is shown on each grain boundary in Fig. 4.1a, b. Furthermore, all grain boundaries have asymmetrical structures as shown by the thick line in Fig. 4.1d; hence, the deviation angles $\phi$ from a symmetrical grain-boundary plane shown by the broken line in Fig. 4.1d for each boundary are provided in parentheses below the $\theta$ values in Fig. 4.1a and b. We also consider the minimum angle $\psi$, shown in Fig. 4.1d, between the two slip planes in neighboring grains for understanding the resistance to a slip transition of dislocations through grain boundaries, and each value is shown in parentheses below the misorientation angle $\theta$ in Fig. 4.1c.

Each model consists of atoms ranging between 1.1 and 3.2 million in dependence on $d$. For each model with $d$ ranging from 5 to 80 nm, a tensile deformation is caused in the $Z$ direction at a strain rate of $8 \times 10^8$ 1/s, whereas the deformation in the $Y$ direction is produced by maintaining the stress $\tau_Y$ at zero by the Parrinello–Rahman method [18]. We consider three analysis temperatures, 100, 300 and 500 K, for model 1a, and 300 K for model 1b. The local face-centered-cubic (fcc) and hexagonal close-packed (hcp) crystal structures in addition to the defect atomic structures are classified using common neighbor analysis (CNA) [19].

## 4.2.2 Results and Discussion

### 4.2.2.1 Grain-Boundary Properties

Figure 4.2 depicts the relationship between the grain-boundary energy and the grain-boundary misorientation angle $\theta$ for asymmetrical tilt grain boundaries in model 1a and model 1b. The asymmetrical grain-boundary energy is calculated by performing simulated annealing at 0.1 K to eliminate the entropy effect for models 1a and 1b with $d = 60$ nm. Values of the equilibrium $\langle 110 \rangle$ symmetrical grain boundaries are also shown in Fig. 4.2 in order to consider the deviation angle $\phi$ influence on the grain-boundary energies. In order to determine the most stable symmetrical grain-boundary structure for $\theta$, we prepare a large number of initial atomic configurations, while taking into account the microscopic degrees of freedom of each grain [20]. The grain-boundary energies for asymmetrical boundaries in models 1a and 1b show higher values than those for symmetrical boundaries. In addition, the asymmetrical grain boundaries with the same $\theta$ show different energy values. Figure 4.3 shows atomic configurations of asymmetrical grain boundaries with $\theta = 67.5$ and $112.5°$ for different values of $\phi$ in models 1a and 1b. Dark, medium, and light gray atoms represent atoms in their neighboring structures with hcp, defects, and fcc structures, respectively. Asymmetrical grain boundary with $\theta = 112.5°$ is close to the twin interface of $\Sigma 3$ with $\theta = 109.5°$. Based on the geometry of grain-boundary structures, the grain-boundary structure with $\theta = 112.5°$ and $\phi = 90°$ is identical to the structure with $\theta = 67.5°$ and $\phi = 0°$. The proportion of twin boundaries in asymmetrical boundaries with $\theta = 112.5°$ clearly decreases with $\phi$. Figure 4.4 shows the deviation angle $\phi$ dependence of grain-boundary energies for $\theta = 22.5°$ and $\theta = 112.5°$. For asymmetrical grain boundaries with $\theta = 112.5°$ containing twin interfaces, grain-boundary energies linearly increases with $\phi$, while for $\theta = 22.5°$, the deviation angle $\phi$ does not affect the grain-boundary energies. Therefore, the grain-boundary energies and structures

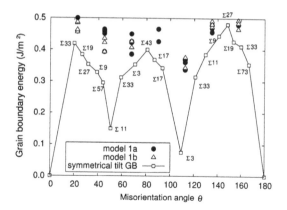

**Fig. 4.2** Grain-boundary energy versus misorientation angle $\theta$ for the $\langle 110 \rangle$ symmetrical and asymmetrical tilt grain boundaries observed in models 1a and 1b with $d = 60$ nm

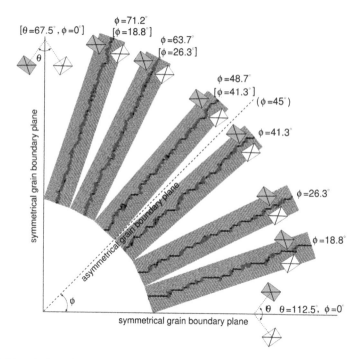

**Fig. 4.3** The $\phi$ dependence of atomic configurations of asymmetric grain boundaries with $\theta = 67.5$ and $112.5°$

both depend on $\phi$ as well as $\theta$, especially in a case where a symmetrical grain-boundary structure is close to the twin interface of $\Sigma 3$ with $\theta = 109.5°$. In addition, Fig. 4.5 shows the relationship between the grain-boundary energies and grain-boundary free volumes for models 1a and 1b. The average energy is 400 mJ/m$^2$ for model 1a and 451 mJ/m$^2$ for model 1b, and the average grain-boundary free volume

**Fig. 4.4** Grain-boundary
energy versus $\phi$

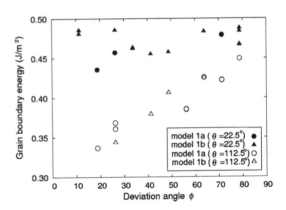

of model 1b is 20% larger than that of model 1a. Hence, we find that each model
contains different average grain-boundary properties in addition to the different distributions of the grain-boundary misorientation angle $\theta$, although these models
consist of the same grains A–H.

Figure 4.6 shows the relationship between the proportion of the grain-boundary
region in simulated nanocrystalline metals ($f_{gb}^{MD}$) and the grain size. The proportion
of the grain-boundary regions dramatically increases with the decrease in the grain
size $d$. The broken curves represent $f_{gb}^{ideal}$, which is estimated by an ideal model
having a grain-boundary thickness of 0.54 nm [21]. The values of $f_{gb}^{MD}$ and $f_{gb}^{ideal}$
are almost identical; therefore, we can infer that the grain-boundary thickness in
both the models does not depend on $d$ and that the grain-boundary thickness
$b$ cannot be neglected when $d$ decreases. The twin boundary is composed of atoms
in the hcp structure as shown in Figs. 4.3 and 4.6. Hence, the proportion of the hcp
structure in the grain boundaries in these quasi 2D models with $\langle 110 \rangle$ asymmetrical tilt grain boundaries is approximately 13% and the value is much higher than
that of three-dimensional (3D) models, i.e., 1.8%, as reported [22]. Therefore, the
analysis models used in this study contain a larger number of twin interfaces than
the polycrystalline aluminum with 3D grain shapes. The reason why the proportion
of the hcp structure in model 1b containing two grain boundaries with $\theta = 112.5°$
is slight larger than that in model 1a containing three grain boundaries with
$\theta = 112.5°$ is that the hcp atoms also exist in grain boundaries with $\theta = 22.5°$.
It has been reported that the grain-boundary structure of $\Sigma 33$ with $\theta = 20.05°$ can
be expressed by the combination of A and B types of structural units for the high
stacking fault energy and A, C, and D types of structural units for the low stacking
fault energy [20]. Here, D type of structural unit is composed of stacking faults.
Although we consider the polycrystalline aluminum with the high stacking fault
energy in this study, asymmetrical grain-boundary structures in models 1a and 1b
do not take the most stable structure after relaxation calculations. Therefore, the
grain boundaries with $\theta = 22.5°$ in a metastable contain the hcp atoms.

**Fig. 4.5** Grain-boundary energy versus grain-boundary free volume

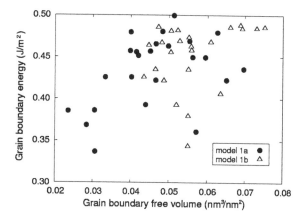

**Fig. 4.6** Proportion of grain-boundary regions in models 1a and 1b. Broken line represents the analytical proportion of grain-boundary region estimated by an ideal model with a grain-boundary thickness of 0.54 nm

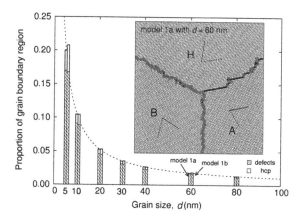

#### 4.2.2.2 Grain-Size Dependence of the Strength

Figure 4.7a shows the stress–strain curves in the case of model 1a at 300 K. In these simulations, nanocrystalline models deform under the following severe conditions: tensile deformation speed is very high, none of the grains contain any defect in the initial state, and crystal slips have geometrical restrictions, increasing the stress to values higher than those in the experiments; further, the flow stress becomes oscillating as the initial grain size becomes large as shown in Fig. 4.7 because the total number of grains in models with a large grain size is less [21]. However, the dependence of the grain size $d$ on the flow stress $\sigma_f$ beyond the peak stress values shows a transition from grain-size hardening to grain-size softening. In this study, the average flow stress $\sigma_f$, estimated by averaging the tensile stress after a strain $\Delta = 0.1$, is regarded as the material strength. Figure 4.7b shows the relationship between the flow stress $\sigma_f$ and the inverse square root of the grain size $d^{-1/2}$ of model 1a at 100, 300, and 500 K. We observe that a transition from grain-size hardening to grain-size softening occurs in both the models: the flow stress increases as the grain

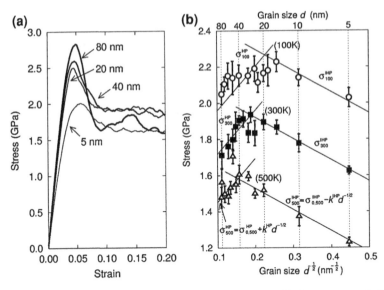

**Fig. 4.7** **a** Stress–strain curves of model 1a at 300 K. **b** Relationship between $\sigma_f$ and $d^{-1/2}$ of model 1a at 100, 300, and 500 K

size decreases to about 20–40 nm, illustrating the Hall–Petch effect [23, 24]. On the other hand, grain-size softening can be observed as the grain size decreases below about 20–40 nm, illustrating the inverse Hall–Petch effect [25, 26]. The solid lines in Fig. 4.7b represent the following relations: $\sigma_T^{HP} = \sigma_{0,T}^{HP} + k^{HP}d^{-1/2}$ for the conventional Hall–Petch region and $\sigma_T^{IHP} = \sigma_{0,T}^{IHP} - k^{IHP}d^{-1/2}$ for the inverse Hall–Petch region. When we assume that $k^{HP}$ and $k^{IHP}$ are intrinsic values independent of temperature, $k^{HP}$ is about 0.076 MN/m$^{3/2}$ and the $k^{IHP}$ is about half of the $k^{HP}$. Here, $k^{HP}$ obtained by atomic simulations takes within the range of $k^{HP}$ reported in actual aluminum [27]. The temperature dependence of the inverse Hall–Petch region is greater than that of the Hall–Petch region. As a result, thermally activated processes are more dominant to the inverse Hall–Petch region than the Hall–Petch region. In the Hall–Petch region, intragranular deformation caused by crystal slips may be dominant; therefore, the grain boundary acts as an obstacle in the path of the dislocation motion. On the other hand, in the inverse Hall–Petch region, intergranular deformation caused by grain-boundary-mediated plasticity, such as grain-boundary sliding and migration may be dominant. The grain size for the maximum strength increases with temperature because of the difference of the temperature dependence of intragranular and intergranular deformation. Consequently, the optimum grain size is not an intrinsic value of a material.

In order to understand the deformation mechanism in the conventional and the inverse Hall–Petch regions, atomic configurations under deformation are examined in detail. First, we consider the inverse Hall–Petch region. Figure 4.8a shows the atomic configuration at a strain of 0.12 in the 5 nm at 300 K, in which grain-size

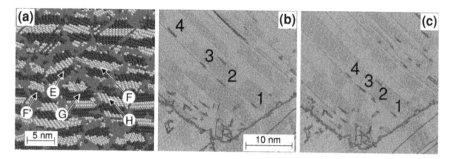

**Fig. 4.8 a** Intergranular deformation of nanocrystalline Al in 5 nm grain at 300 K when $\varepsilon = 0.12$. **b, c** A pile-up of dislocations 1–4 in grain B in the vicinity of the grain boundary with $\theta = 67.5°$ for $d = 80$ nm when $\varepsilon = 0.164$ and $0.176$

softening can be observed. The dark gray atoms represent the local defects and the hcp structure, whereas the medium gray and light gray atoms represent the local fcc structure. The medium gray and light gray stripes are marked on the initial atomic configurations. No dislocation core exists in the grains. Steps in the stripes at the grain boundary between grain H and grain G are visible; therefore, grain-boundary sliding occurs in the grain boundary. In general, grain-boundary sliding produces geometrical misfits at triple junctions such as cracks, which is one of the causes of intergranular fracture. Therefore, accommodation mechanisms for misfits are required to obtain large elongation. One such mechanism would be a grain-boundary diffusion creep mechanism [28] and another possibility is a grain rotation mechanism proposed by theoretical analyses [29, 30]. As proposed in the theoretical analyses, grain rotations of grain E and grain F can be observed in this atomic simulation, and grain rotations can accommodate geometric misfits caused by the grain-boundary sliding. Therefore, the dominant deformation mechanism in the grain-size softening regions is intergranular deformation.

Next, we consider the conventional Hall–Petch region. Figures 4.8b and 4.9c show atomic configurations of 80 nm grains. Several dislocations can move simultaneously in grain B. A pile-up of dislocations 1–4 toward the grain boundary among grain B and grain D can be observed in grain B, but the respective crystal glide planes of dislocations 1–4 are not identical. A pile-up of dislocations facilitates activation of intragranular deformation in adjacent grains, namely the propagation of plastic deformation because the stress concentration that is attributable to a pile-up of dislocations near the grain boundaries occurs, illustrating the Hall–Petch mechanism. Therefore, the dominant deformation mechanism in the grain-size hardening regions is dislocation movement.

### 4.2.2.3 Influence of Grain-Boundary Structures on the Strength

Next, we consider the influence of grain-boundary structures on the strength of polycrystalline metals. Figure 4.9 shows the relationship between the flow stress $\sigma_f$ and the inverse square root of the grain size $d^{-1/2}$ at 300 K for models 1a and 1b

**Fig. 4.9** Relationship between $\sigma_f$ and $d^{-1/2}$ of models 1a and 1b at 300 K

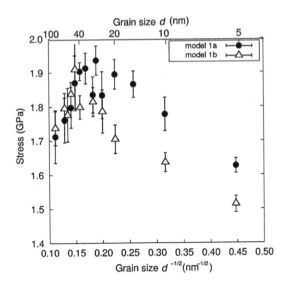

with different grain-boundary characteristics. Although the analysis conditions for each model are the same, the deformation resistances for the Hall–Petch and inverse Hall–Petch regions are different. The deformation resistance in the grain-size hardening region in model 1b is larger than that in model 1a; however, this tendency is reversed in the grain-size softening region. The result indicates that the role of the grain-boundary changes as the grain size decreases, and the intergranular and intragranular deformations may strongly depend on the grain-boundary structure defined by $\theta$ and $\phi$. For the Hall–Petch regions, the dislocation source ability of grain boundaries and the minimum angle $\psi$ defined in Fig. 4.1d may be a key factor in determining the resistance to intragranular deformation. Figure 4.10 shows the atomic configurations when dislocations are generated from triple junctions and grain boundaries in models 1a and 1b with $d = 60$ nm at 300 K, respectively. Because of stress concentrations caused by the elastic-strain incompatibility near triple junctions, some dislocations are emitted from the triple junctions. Although some papers investigating triple junction structures by using atomic simulations [31], the relationship between the structures and mechanical properties have not been cleared yet due to a large number of required degrees of freedom to express the structures. However, a large number of researches about dislocation emissions from grain boundaries in nanocrystalline models [6, 7, 21, 32, 33] and tilt grain boundaries in bicrystal models [34–37] are reported by using atomic simulations. For example, it has been shown for asymmetrical grain boundaries that ledges in a grain-boundary plane, which accommodate a deviation from the symmetrical boundary plane, are effective dislocation sources [36], and $E$ structural units in $\langle 110 \rangle$ tilt grain boundaries also become productive sites for generating dislocations [35]. As you can see in Fig. 4.10, dislocation emissions occur from asymmetrical grain boundaries with $\theta = 112.5°$ as predicted by the

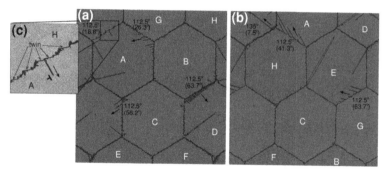

**Fig. 4.10** Dislocation emissions from grain boundaries in **a** model 1a and **b** model 1b with $d = 60$ nm at 300 K

reported bicrystal models [36]. Therefore, we can confirm that asymmetrical grain boundaries close to $\Sigma 3$ can emit dislocations easier than other boundaries in polycrystalline models, and model 1a have a larger number of effective dislocation sources than model 1b as shown in Fig. 4.1. Model 1a also has a larger number of grain boundaries with small $\psi$ than model 1b as shown in Fig. 4.1c; hence, the propagation of intragranular deformation is more difficult to occur in model 1b than in model 1a, as shown in Fig. 4.9. These results strongly affect the resistance of intragranular deformation. On the other hand, it has been reported that the grain-boundary-mediated plasticity can easily occur when grain boundaries have high grain-boundary energies and large free volumes [38]. The average grain-boundary energy and free volume in model 1b is larger than that in model 1a; therefore, the resistance of intergranular deformation for model 1b is smaller than that for model 1a, as shown in Fig. 4.9.

Consequently, the grain-boundary structures and their properties related to intergranular and intragranular deformations strongly govern polycrystalline metals when the grain size decreases to the order of several hundred nanometers, and hence, it is very important for us to obtain a clear picture of the roles of grain boundaries in ultrafine-grained metals.

## 4.3 Influence of Grain-Boundary Structure on Dislocation Transmissions

In the previous section, we investigated the roles of grain boundaries in the strength of ultrafine-grained metals by using molecular dynamics simulations, and we considered that the minimum angle $\psi$ between the two slip planes in neighboring grains affect the resistance to intragranular deformation. In this section, we perform a further detailed examination of the mechanism of the interactions between dislocations and tilt grain boundaries considering the grain-boundary structures by using the quasicontinuum method, which is one of a multi-scale method that combines the atomic and continuum regions.

## 4.3.1 Models and Methodologies

### 4.3.1.1 Quasicontinuum Method

The basic idea of the quasicontinuum method [13] is conceptually simple. To perform an atomic simulation efficiently and to save computational resources, a continuum approximation in which atomic deformation-gradient fields are small is adopted, assuming that the continuum method provides almost the same result as a full atomistic simulation. However, it is actually very difficult to define the energy of the interface region in such a way that the coupling of the atomic description (nonlocal atoms) and the continuum description (local atoms) is completely seamless: unphysical forces (ghost forces) appear at interfaces because the fact that the interaction range of a nonlocal atom is different from that of a local atom [39], leading, for example, to serious problems with energy conservation during molecular dynamics simulations. In this study, we demonstrate a seamless coupling of the two descriptions by introducing a buffer layer of a new type of atom called a quasi nonlocal atom.

Figure 4.11a shows a usual QC model (model $QC^1$) near a vacancy. Here, open circles near the vacancy are referred to as nonlocal [39] because the energy depends on the positions of all the atoms within a certain cutoff distance $R_c$ from the atoms. Solid circles far from the vacancy indicate the so-called local atoms forming the corners of the triangulation of the region with a slowly varying deformation gradient. When the force acting on atom $\alpha$ in the model $QC^1$ is compared to that acting on an atom in the full atomistic model, we can confirm that these forces are not equivalent: a ghost force appears at the interface between the nonlocal and local regions. To solve this problem, we introduce a new type of atom, named a quasi-nonlocal atom, between the nonlocal and local regions so that no nonlocal atom interacts with any local atom [15]. The concept of a quasi-nonlocal atom is very simple. A quasi-nonlocal atom can feel first nearest neighbor atoms and all nonlocal atoms within the cutoff distance $R_c$. Figure 4.11b shows the concept of the improved QC model (model $QC^2$). Double circles correspond to quasi-nonlocal atoms.

The potential energy of a quasi-nonlocal atom is calculated as if it was a non-local atom with one important difference: while calculating the energy of a quasi-nonlocal atom, the positions of only the nearest neighbor atoms and nonlocal atoms within the cutoff distance are used. On the local side of the interface, the distance vectors to the nearest neighbor atoms are used to extrapolate the positions of the second nearest neighbors, third nearest neighbors, and so on, as illustrated in Fig. 4.11c. Large open circles and double circles around the quasi-nonlocal atom $\alpha$ represent the nearest neighbor atoms; gray circles represent the extrapolated

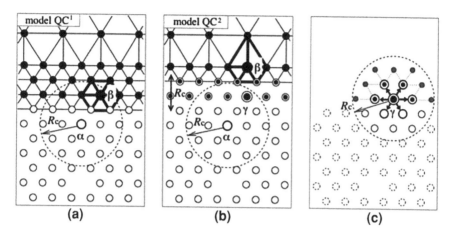

**Fig. 4.11** Coupling of atomic and continuum model with one vacancy. *Open*, *closed* and *double circles* show nonlocal, local, and quasi-nonlocal atoms, respectively

neighbor atoms; and broken circles represent the nonlocal atoms not required for calculating the potential energy of the quasi-nonlocal atom $\alpha$. If the position of an extrapolated neighbor atom and that of a nonlocal atom are almost the same, the nonlocal atom's position is used. By using these extrapolated neighbor atoms, we have enough information to calculate the potential energy of a quasi-nonlocal atom in the same manner as that for a nonlocal atom. Consequently, a quasi-nonlocal atom acts like a nonlocal atom on the nonlocal side of the interface and like a local atom on the local side of the interface.

To avoid ghost forces, the range of the potential must be limited to first neighbors of the first neighbors. In this manner, if atom 1 extrapolates the position of atom 2 by using common nearest neighbors, atom 2 extrapolates the position of atom 1 by using the same neighbors, and the situation is symmetric. The extrapolation limit that no ghost force appears concerns only the positions of atoms that are first nearest neighbors to a first nearest neighbor of a quasi-nonlocal atom: a quasi-nonlocal atom named $\gamma^1$ extrapolates a neighbor atom at position 2 by using its first neighbor atoms, and similarly, a quasi-nonlocal atom $\gamma^2$ at position 2 extrapolates a neighbor atom at position 1 by using the common first neighbor atoms. Therefore, no asymmetry is introduced. The limit is fourth and third nearest neighbors for fcc and bcc lattices, respectively. If the range of the potential goes beyond this limit, ghost forces can no longer be avoided. As an example, let us observe the fcc structure shown in Fig. 4.12. Full circles express nearest neighbors to the black atom $\gamma$ and medium, large, and small circles represent atoms in the same plane, one plane above, or one plane below the black atom $\gamma$, respectively. Dashed circles correspond to first neighbors to the first neighbor $\eta^a$ of the atom $\gamma$. We can extrapolate neighbors by using the 12 different vectors $d_\eta = r_\eta - r_\gamma$ as follows: $r_{2nd} = r_\eta + 1/2(d_a + d_f + d_j + d_k)$, $r_{3rd} = r_\gamma + d_a + d_j$, and $r_{4th} = r_\gamma + 2d_a$.

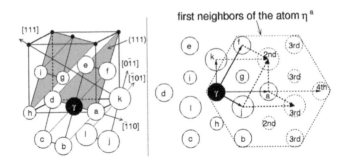

**Fig. 4.12** First nearest neighbors to the atom $\gamma$ and superposed atomic configurations on (111) planes for f.c.c. structures

### 4.3.1.2 Analysis Models and Conditions

Figure 4.13a shows the schematic of the analysis model used for investigating the interaction mechanism between the edge dislocations and the aluminum $\langle 112 \rangle$ asymmetrical tilt grain boundaries through quasicontinuum simulations [40]. The embedded atom method for aluminum proposed by Mishin et al. [16] is adopted for atomic interactions. The $\langle 112 \rangle$ tilt boundaries are adopted because periodic boundary conditions can be applied to the $\langle 112 \rangle$ tilt grain boundaries with a tilt axis parallel to the lattice edge dislocation line and the defect geometries. The analysis model comprises two crystal grains A and B, and the dislocation source in grain A is a crack. In this study, we intend to express the movement of dislocations from the crack to the grain boundary and the interactions between the dislocations and the grain boundary. Therefore, the region near the dislocation slip plane in grain A and that near the grain boundary between grains A and B have full atomistic resolution. On the other hand, the regions distant from the slip plane or the grain boundary are divided into finite elements. Therefore, the mechanics of the atoms in an element are determined by the positions of the node atoms. The enlarged picture in Fig. 4.13a shows the atomic arrangements near the coupling interface between the atomistic and continuum regions. The solid, open, and double circles represent the *nonlocal* atoms corresponding to the atomistic region, the *local* atoms corresponding to the continuum regions, and the *quasi-nonlocal* atoms corresponding to the buffer layers that seamlessly couple the atomistic and continuum regions, respectively, as mentioned in the previous paragraph [15]. Periodic boundary conditions are adopted along the $X$ and $Z$ directions.

In this study, we prepare two models with different grain-boundary structures. To generate edge dislocations from the crack in grain A under shear deformation, as shown in Fig. 4.13a, the crystal orientations along the $x^A$, $y^A$, and $z^A$ directions of grain A in both models are set to [11-2], [-110], and [111], respectively. The crystal orientation along the $x^B$ direction of grain B is also set to [11-2]; therefore, the grain-boundary structures in these models are controlled by the rotational angle $\theta$ about the $X$ direction. Two small-angle tilt grain boundaries with misorientation angles of 13 and 30° are considered; these two boundaries characterize model 2a shown in

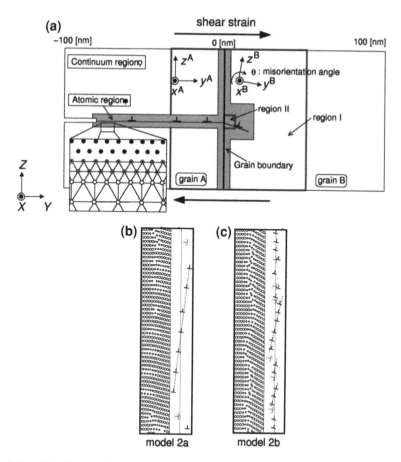

**Fig. 4.13  a** Coupling model to investigate the interaction between edge dislocations and tilt grain boundaries in quasicontinuum method. **b, c** Grain-boundary structures in model 2a with $\theta = 13°$ and $\phi = 7°$ and in model 2b with $\theta = 29.5°$ and $\phi = 14°$, respectively

Fig. 4.13b and model 2b shown in Fig. 4.13c, respectively. For each model, the dimensions along the $X$, $Y$, and $Z$ directions are approximately 24, 200, and 55 nm, respectively. The distance between the crack tip and the grain boundary is approximately 40 nm. The sum of the nonlocal, local, and quasi-nonlocal atoms in each model is approximately 1.5 million. If the analysis model is expressed only by nonlocal atoms, the total number of atoms is approximately 16 million. Consequently, the degree of freedom in this QC model is one-tenth of that in a full atomistic model.

To simulate the interactions between the incoming dislocations and the tilt grain boundaries, the shear strain increment $\Delta\gamma_{ZY}$ is repeatedly applied to the analysis models. The energy of the analysis model for each $\Delta\gamma_{ZY}$ is minimized via the conjugate gradient method; no thermally activated process is considered in these simulations. The value of $\Delta\gamma_{ZY}$ is 0.002. Under shear deformation, the $Z$-directional displacement of the surface nodes along the $Y$ direction is fixed.

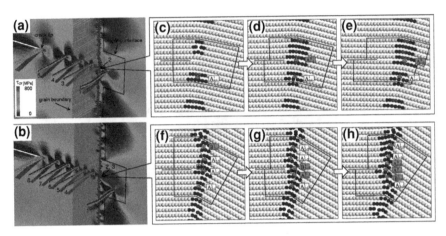

**Fig. 4.14** Interactions between lattice edge dislocations from the crack tip and asymmetrical tilt grain boundaries under shear deformation. **a** Model 2a: $\theta = 13.0°$, $\phi = 7°$, $\gamma_{ZY} = 0.024$; **b** model 2b: $\theta = 29.5°$, $\phi = 14°$, $\gamma_{ZY} = 0.034$. Atomic configurations of the tilt grain boundaries in model 2a and 2b under shear deformation: **c** $\gamma_{ZY} = 0.002$, **d** $\gamma_{ZY} = 0.012$, **e** $\gamma_{ZY} = 0.024$, **f** $\gamma_{ZY} = 0.002$, **g** $\gamma_{ZY} = 0.012$, and **h** $\gamma_{ZY} = 0.028$

## 4.3.2 Results and Discussion

### 4.3.2.1 Interactions Between Dislocations and Grain Boundaries

Figure 4.14a, b shows the final atomic configurations in each model under the shear deformation applied in this study. To observe the defect structures easily, the atoms in the local fcc structure are not shown in Fig. 4.14a, b. The atoms in the local hcp and defect structures are depicted in brown and gray, respectively. The defect atoms in the grain boundaries are shown as transparent circles. The dislocations resulting from the crack tip are numbered according to their order of appearance. The distributions of $\tau_{ZY}$ are also shown in each background. The black lines show the coupling interface between the atomistic and continuum regions in grain B, and the seamless distribution of $\tau_{ZY}$ can also be observed. The dislocation transmission in model 2a occurs when $\gamma_{ZY}$ attains a value of 0.024 and the grain boundary absorbs the second incoming dislocation, as shown in Fig. 4.14a. Similarly, the dislocation transmission in model 2b occurs when $\gamma_{ZY}$ attains the value of 0.034 and two incoming dislocations are absorbed by the grain boundary, as shown in Fig. 4.14b.

To elucidate the accommodation of the incoming dislocations in the grain boundaries, Fig. 4.14c–h show a series of atomic configurations near the grain boundaries, where the incoming dislocations are absorbed. Analytically, two sets of uniformly spaced edge dislocations are required to construct an asymmetric low-angle tilt grain boundary [41]. The Burgers vectors of these grain-boundary dislocations, $b_1$ and $b_2$, are perpendicular to each other. In model 2a, one vacancy and one primary intrinsic grain-boundary dislocation (IGBD) are found in the

Burgers circuit, as shown in Fig. 4.14c. No secondary IGBD is observed near the slip plane; hence, the grain boundary around the slip plane can be considered to be symmetrical. In model 2b, as shown in Fig. 4.14f, three primary IGBDs and one secondary IGBD are found in the Burgers circuit. Hence, the effect of the secondary IGBD on the accommodation of the extrinsic grain-boundary dislocations (EGBDs) can be investigated by comparing the results of these two models.

When $\gamma_{ZY}$ attains a value of 0.012 in model 2a, the incoming dislocation is absorbed at the site between the vacancy and the IGBD $A^I_{1-1}$ as the EGBD $A^E_{1-1}$, as shown in Fig. 4.14d. Subsequent dislocations from the crack tip pile-up behind $A^E_{1-1}$ with an increase in $\gamma_{ZY}$, as shown in Fig. 4.14a. Finally, $A^E_{1-1}$ is ejected from the grain boundary when $\gamma_{ZY} = 0.024$ and the next incoming dislocation is absorbed as $A^E_{1-2}$ at the same site, as shown in Fig. 4.14e. Consequently, this grain boundary can accommodate only one EGBD in the slip plane of the incoming dislocations.

In the equilibrium state in model 2b, as shown in Fig. 4.14f, the IGBD $A^I_{2-2}$ already exists in the slip plane of the incoming dislocation. Therefore, we can easily infer that the first incoming dislocation will pile-up behind $A^I_{2-2}$, as in the case of model 2a. However, the first incoming dislocation does not pile-up but is absorbed by the grain boundary as $A^E_{2-1}$, as shown in Fig. 4.14g. This phenomenon is due to the sliding of the secondary IGBD $B^I_{2-1}$ along the grain boundary and the ejection of the atomic group in the red (dark gray) box, which corresponding to the magnitude of the Burgers vector of the incoming edge dislocation, from its atomic plane. Consequently, the dislocations $A^I_{2-2}$, $A^I_{2-3}$, and $A^E_{2-1}$ change their atomic planes; the dislocation climb ups and climb downs occur without the diffusion process. As $\gamma_{ZY}$ increases, the incoming dislocations pile-up behind $B^I_{2-1}$. When attains the value of 0.028, $B^I_{2-1}$ slides once again along the direction of the arrow, and the second incoming dislocation is simultaneously absorbed as $A^E_{2-2}$ with a climb up caused by the sliding of $B^I_{2-1}$. There is no secondary IGBD in the upper region of the slip plane, as observed in Fig. 4.14h; therefore, the EGBDs are not accommodated at the grain boundary and the incoming dislocations pile-up behind $A^E_{2-1}$. Finally, $A^E_{2-1}$ is ejected from the grain boundary by the dislocation pile-up when $\gamma_{ZY}$ attains the value of 0.034. Note that the secondary IGBD $B^I_{2-1}$ does not slide alone under a macroscopic shear stress $\tau_{ZY}$ of 200–400 MPa in this study. Hence, the stress-assisted sliding along the grain-boundary plane of the secondary IGBD with the absorption of the incoming dislocations can strongly affect the accommodation of the EGBDs.

### 4.3.2.2 Stress Concentrations Due to Dislocation Pile-Ups

Figure 4.15 shows the relationship between the macroscopic and microscopic shear stresses ($\tau^{Mac}_{ZY}$ and $\tau^{Mic}_{ZY}$) and the shear strain $\gamma_{ZY}$ in order to investigate the influence of the dislocation pile-up on the stress concentration. The macroscopic and microscopic stresses are evaluated in region I: $-30$ nm $\leq Y \leq 50$ nm, and in region II: 0 nm $\leq Y \leq 5$ nm and $-2.5$ nm $\leq Z \leq 2.5$ nm, respectively (as shown

**Fig. 4.15** Relationship
between macroscopic and
microscopic shear stresses
and $\gamma_{ZY}$

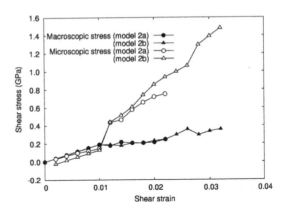

in Fig. 4.13a). The slopes of $\tau_{ZY}^{Mac}$ and $\tau_{ZY}^{Mic}$ are almost identical for both the models before the first dislocation emission from the crack tip because the anisotropy factor, $2C_{44}/(C_{11}-C_{12})$, is very close to one (approximately 1.25) for the adopted atomic potential. After the first dislocation emission from the crack tip when $\gamma_{zy} = 0.012$, the slopes of $\tau_{ZY}^{Mac}$ decrease in comparison with those under elastic deformation. We also clearly observe that the increment rate of $\tau_{ZY}^{Mic}$ is larger than that of $\tau_{ZY}^{Mac}$ because of the dislocation pile-up and that the increment rate of $\tau_{ZY}^{Mic}$ does not exhibit a strong dependence on the grain-boundary structures.

### 4.3.2.3 Critical Glide Force on the Outgoing Edge Dislocation

To investigate the influence of grain-boundary structures on the critical glide force $f_c$ on the edge dislocation to eject from the tilt grain boundaries in models 2a and 2b, the force $f_c$ is directly calculated using the stress field near the dislocation that will be ejected in the QC simulations. $f_c$ is evaluated as the force generated by the average shear stress $\tau_{zByB}^{Mic}$ just before the dislocation ejection. $\tau_{zByB}^{Mic}$ is the microscopic shear stress estimated for the material coordinate system of grain B in region II. $\tau_{zByB}^{Mic}$ is 697 and 811 MPa when $\gamma_{ZY} = 0.022$ in model 2a and $\gamma_{ZY} = 0.032$ in model 2b, respectively. Thus $f_{c, 2a}$ in model 2a and $f_{c, 2b}$ in model 2b can be obtained as follows: $f_{c, 2a} = 0.199$ N/m, $f_{c, 2b} = 0.232$ N/m. Here, the ratio of $f_{c, 2a}$ to $f_{c, 2b}$, ($f_{c, 2a}/f_{c, 2b}$), is approximately 1.17. If dislocations are emitted from a Frank–Read source [42] near the grain boundary with the same length of the segment of the dislocation whose ends are pinned, the critical force $f_c$ shows the same value even though grain-boundary structures are different. However, if dislocations are emitted from the grain boundaries, the critical force $f_c$ exhibits a dependence on the grain-boundary structure. As a result, in ultrafine-grained metals, the effect of the grain-boundary characteristics on the macroscopic mechanical properties should not be ignored, and we can confirm that atomic simulations play an important role in elucidating the mechanical properties of such materials.

## 4.4 Conclusions

In this paper, we have demonstrated two problems in which atomic modeling provides a useful tool in study the roles of grain boundaries in the strength in polycrystalline metals in order to design a high-functional material with high strength. In the first problem, we prepare two polycrystalline models consisting of the same eight grains but having different grain arrangements and investigate the grain-size dependence of the grain-boundary roles by performing tensile deformation tests for the two types of models by using conventional molecular dynamics simulations. We have clearly observed that a transition from grain-size hardening to grain-size softening can be observed as grain size decreases to nanometer order, i.e., the primary role of grain-boundary changes in each region; therefore, the optimum grain size for the strength is not an intrinsic value of a material due to the difference of the temperature and the grain-boundary structure dependence of intragranular and intergranular deformation. In the second problem, we perform a further detailed examination of the mechanism of the interactions between dislocations and tilt grain boundaries considering the grain-boundary structures by using the multiscale atomic method. We have confirmed that the stress concentration due to the dislocation pile-ups near the grain boundary does not show the strong grain-boundary structure dependence, but the critical glide force on the outgoing dislocation from the grain boundary exhibits a dependence on the grain-boundary structure. The two discussed problems show that atomic simulations have the advantage to provide detail continuous atomic-scale information about defect interactions including grain boundaries that cannot be easily obtained by experimental work. To obtain further quantitative results which can be compared with actual experimental results directly, we should expand the limitations of treatable time and space in atomic simulations by adopting new techniques such as the free-end nudged elastic band method [43, 44] and the dynamic quasicontinuum [45] and coupled atomistic/discrete dislocation method [46, 47].

**Acknowledgments** This study was financially supported in part by a Grant-in-Aid for Scientific Research from the Ministry of Education, Culture, Sports, Science and Technology (MEXT), Japan on Priority Areas "Giant Straining Process for Advanced Materials Containing Ultra-High Density Lattice Defects." and also supported in part by a Grant-in-Aid for Scientific Research on Innovative Area, Bulk Nanostructured Metals, through MEXT, Japan (Contract No. 22102007), and these supports are gratefully appreciated.

## References

1. Iwahashi Y, Horita Z, Nemoto M, Wang J, Langdon TG (1996) Principle of equal-channel angular pressing for the processing of ultra-fine grained materials. Scr Mater 35:143–146
2. Saito Y, Utsunomiya H, Tsujia N, Sakaia T (1999) Novel ultra-high straining process for bulk materials development of the accumulative roll-bonding (ARB) process. Acta Mater 47:579–583

3. Valiev RZ, Ivanisenko YV, Rauch EF, Baudelet B (1996) Structure and deformaton behaviour of armco iron subjected to severe plastic deformation. Acta Mater 44:4705–4712
4. Lu L, Shen Y, Chen X, Qian L, Lu K (2004) Ultrahigh strength and high electrical conductivity in copper. Science 304:422–426
5. Van Swygenhoven H, Derlet PM (2001) Grain-boundary sliding in nanocrystalline fcc metals. Phys Rev B 64:224105
6. Schiøtz J, Jacobsen KW (2003) A maximum in the strength of nanocrystalline copper. Science 301:1357–1359
7. Yamakov V, Wolf D, Phillpot SR, Mukherjee AK, Gleiter H (2003) Deformation mechanism crossover and mechanical behaviour in nanocrystalline materials. Phil Mag Lett 83:385–393
8. Wolf D, Yamakov V, Phillpot SR, Mukherjee A, Gleiter H (2005) Deformation of nanocrystalline materials by molecular-dynamics simulation: relationship to experiments? Acta Mater 53:1–40
9. Cahn JW, Mishin Y, Suzuki A (2006) Coupling grain boundary motion to shear deformation. Acta Mater 54:4953–4975
10. Broughton JQ, Abraham FF, Bernstein N, Kaxiras E (1999) Concurrent coupling of length scales methodology and application. Phys Rev B 60:2391–2403
11. Shiari B, Miller RE, Curtin WA (2005) Coupled atomistic/discrete dislocation simulations of nanoindentation at finite temperature. J Eng Mater Tech 127:358–368
12. Miller RE, Tadmor EB (2009) A unified framework and performance benchmark of fourteen multiscale atomistic/continuum coupling methods. Model Simul Mater Sci Eng 17:053001
13. Tadmor EB, Phillips R, Ortiz M (1996) Mixed atomistic and continuum models of deformation in solids. Langmuir 12:4529–4534
14. Miller RE, Tadmor EB (2002) The quasicontinuum method: overview, applications and current directions. J Comput Aided Mater Des 9:203–239
15. Shimokawa T, Mortensen JJ, Schiøtz J, Jacobsen KW (2004) Matching conditions in the quasicontinuum method: removal of the error introduced at the interface between the coarse grained and fully atomistic region. Phys Rev B 69:214104
16. Mishin Y, Farkas D, Mehl MJ, Papaconstantopoulos DA (1999) Interatomic potentials for monoatomic metals from experimental data and ab initio calculations. Phys Rev B 59:3393–3407
17. Yamakov V, Wolf D, Salazar M, Phillpot SR, Gleiter H (2001) Length-scale effects in the nucleation of extended dislocations in nanocrystalline Al by molecular-dynamics simulation. Acta Mater 49:2713–2722
18. Parrinello M, Rahman A (1981) Polymorphic transitions in single crystals: a new molecular dynamics method. J Appl Phys 52:7182–7190
19. Jónsson H, Andersen HC (1988) Icosahedral ordering in the Lennard-Jones liquidand glass. Phys Rev Lett 60:2295–2298
20. Rittner JD, Seidman DN (1996) ⟨110⟩ symmetric tilt grain-boundary structures in fcc metals with low stacking-fault energies. Phys Rev B 54:6999–7015
21. Shimokawa T, Nakatani A, Kitagawa H (2005) Grain-size dependence of relationship between intergranular and intragranular deformation of nanocrystalline al by molecular dynamics simulations. Phys Rev B 71:224110
22. Shimokawa T, Nakatani A, Kitagawa H (2004) Mechanical properties depending on grain sizes of fcc nanocrystalline metals by using molecular dynamics simulation (investigation of stacking fault energy's influence). JSME Int J A 47:83–91
23. Hall EO (1951) The deformation and ageing of mild steel: III discussion of results. Proc Phys Soc B 64:747–753
24. Petch NJ (1953) The cleavage strength of polycrystals. J Iron Steel Inst 174:25
25. Chokshi AH, Rosen A, Karch J, Gleiter H (1989) On the validity of the hall-petch relationship in nanocrystalline materials. Scripta Met 23:1679–1684
26. Fougere GE, Weertman JR, Siegel RW, Kim S (1992) Grain-size dependent hardening and softening of nanocrystalline cu and pd. Scripta Met Mater 26:1879–1881

27. Wyrzykowski JW, Grabski MW (1986) The Hall-Petch relation in aluminum and its dependence on the grain boundary structure. Phil Mag A 53:505–520
28. Yamakov V, Wolf D, Phillpot SR, Gleiter H (2003) Dislocation-dislocation and dislocation-twin reactions in nanocrystalline al by molecular dynamics simulation. Acta Mater 51:1971–1987
29. Gutkin MY, Ovid'ko IA, Skiba NV (2003) Crossover from grain boundary sliding to rotational deformation in nanocrystalline materials. Acta Mater 51:4059–4071
30. Ovid'ko LA (2002) Deformation of nanostructures. Science 295:2386
31. Frolov T, Mishin Y (2009) Molecular dynamics modeling of self-diffusion along a triple junction. Phys Rev B 79:174110
32. Van Swygenhoven H (2002) Grain boundaries and dislocations. Science 296:66–67
33. Frøseth A, Van Swygenhoven H, Derlet PM (2004) The influence of twins on the mechanical properties of nc-al. Acta Mater 52:2251–2258
34. Spearot DE, Tschopp MA, Jacob KI, McDowell DL (2007) Tensile strength of $\langle 100 \rangle$ and $\langle 110 \rangle$ tilt bicrystal copper interfaces. Acta Mater 55:705–714
35. Tschopp M, Tucker G, McDowell D (2007) Structure and free volume of $\langle 110 \rangle$ symmetrictilt grain boundaries with the E structural unit. Acta Mater 55:3959–3969
36. Tschopp M, McDowell D (2008) Dislocation nucleation in $\Sigma 3$ asymmetric tilt grain boundaries. Int J Plas 24:191–217
37. Shimokawa T (2010) Asymmetric ability of grain boundaries to generate dislocations under tensile or compressive loadings. Phys Rev B 82:174122
38. Monzen R, Suzuki T (1996) Nanometre-scale grain-boundary sliding in copper bicrystals with [001] twist boundaries. Philo Mag Lett 74:9–15
39. Shenoy VB, Miller R, Tadmor EB, Rodney D, Phillips R, Ortiz M (1999) An adaptive finite element approach to atomic-scale mechanics the quasicontinuum method. J Mech Phys Solids 47:611–642
40. Shimokawa T, Kinari T, Shintaku S (2007) Interaction mechanism between edge dislocations and asymmetrical tilt grain boundaries investigated via quasicontinuum simulations. Phys Rev B 75:144108
41. Read WT, Shockley W (1950) Dislocation models of crystal grain boundaries. Phys Rev 78:275–289
42. Hirth JP, Lothe J (1968) Theory of dislocations, 2nd edn. McGraw-Hill, NewYork
43. Zhu T, Li J, Samanta A, Kim HG, Suresh S (2007) Interfacial plasticity governs strain rate sensitivity and ductility in nanostructured metals. Proc Natl Acad Sci 104:3031–3036
44. Zhu T, Li J, Samanta A, Leach A, Gall K (2008) Temperature and strain-rate dependence of surface dislocation nucleation. Phys Rev Lett 100:025502
45. Dupuy L, Tadmor E, Miller R, Phillips R (2005) Finite-temperature quasi-continuum: molecular dynamics without all the atoms. Phys Rev Lett 95:1–060202
46. Qu S, Shastry V, Curtin WA, Miller RE (2005) A finite-temperature dynamic coupled atomistic/discrete dislocation method. Model Simul Mater Sci Eng 13:1101–1118
47. Warner DH, Curtin WA, Qu S (2007) Rate dependenceof crack-tip processes predicts twinning trends in f.c.c. metals. Nat Mater 6:876–881

# Chapter 5
# Semi Emprical Low Cycle Fatigue Crack Growth Analysis of Nanostructure Chip-To-Package Copper Interconnect Using Molecular Simulation

**S. Koh, A. Saxena, Willem Van Driel, G. Q. (Kouchi) Zhang and R. Tummala**

**Abstract** ITRS has predicted that integrated chip (IC) packages will have interconnections with I/O pitch of 90 nm by the year 2018. Lead-based solder materials in flip chip technology will not be able to satisfy the thermal mechanical requirement of these fine pitches. Of all the known interconnect technologies, nanostructure interconnects such as nanocrystalline Cu are the most promising technology to meet the high mechanical reliability and electrical requirements of next generation devices. However, there is a need to fully characterize their fatigue properties. In this research, numerical analysis has been employed to study the semi-elliptical crack growth and shape evolution in nanostructured interconnects subject to uniaxial fatigue loading. The results indicate that nanocrystalline copper is in fact a suitable candidate for ultra-fine pitch interconnects applications. This

S. Koh (✉) · W. Van Driel · G. Q. (Kouchi) Zhang
Delft Institute of Microsystems and Nanoelectronics (Dimes),
Delft University of Technology, Mekelweg 6, 2628 CD Delft,
The Netherlands
e-mail: S.W.koh@tudelft.nl

S. Koh
Materials Innovation Institute (M2i), Mekelweg 2, 2628 CD Delft,
The Netherlands

A. Saxena
Department of Mechanical Engineering, University of Arkansas,
Fayetteville, AR 72701, USA

W. Van Driel · G. Q. (Kouchi) Zhang
Philips Lighting, LightLabs, 5611 BD Eindhoven, The Netherlands

R. Tummala
Department of Materials Science and Engineering,
Georgia Institute of Technology, Atlanta, GA 30332, USA

N. Iwamoto et al. (eds.), *Molecular Modeling and Multiscaling Issues
for Electronic Material Applications*, DOI: 10.1007/978-1-4614-1728-6_5,
© Springer Science+Business Media, LLC 2012

study also predicts that crack growth is a relatively small portion of the total fatigue life of interconnects under LCF conditions. Hence, crack initiation life is the main factor in determining the fatigue life of interconnects.

## 5.1 Introduction

Driven by the need to increase the system functionality and decrease the feature size, the International Technology Roadmap for Semi-conductors (ITRS) has predicted that integrated chip (IC) packages will have interconnections with I/O pitch of 90 nm by the year 2018 [1]. Lead-based solder materials have been used for many decades as interconnections in flip chip technology which will not be able to satisfy the thermal mechanical requirement of these fine pitches [2, 3]. Of all the known interconnect technologies, nanostructure interconnects such as nanocrystalline Cu are the most promising technology to meet the high mechanical reliability and electrical requirements of next generation devices. However, there is a need to fully characterize and understand its failure mechanisms such as cyclic fatigue.

Finite element analysis is one of the most common methods to evaluate the fatigue performance of these interconnects. However, implementing it to study crack evolution during fatigue cycle is a very time consuming process since a new model and analysis is needed for each fatigue cycle which could run up to thousands of cycles before failure. Hence, in order to avoid the vast number of finite element experiments needed to simulate the observed crack growth and shape evolution in experiments; close form solutions for stress intensity factor need to be developed.

In this research, numerical analysis has been employed to study the semi-elliptical crack growth and shape evolution in nanostructured interconnects subject to uniaxial fatigue loading. Molecular dynamics in conjunction to phase mixture model has been employed to determine the elastic modulus of nanocrystalline Cu needed to conduct the crack growth analysis. This method will investigate:

(a) the validity of the fracture mechanics approach in estimating the crack growth life of interconnects.
(b) the relative contribution of the crack growth life to the overall fatigue life of nanostructured Cu interconnects.

## 5.1.1 Proposed Methodology for Crack Growth Analysis

Due to the large strain experienced by the Cu interconnects, the remaining ligament around the semi-elliptical cracks undergoes elastic plastic deformation for microcrystalline Cu interconnects and nanostructured interconnects. Under the circumstances, it is more appropriate to use J-integral as shown in Eq. 5.1 [4].

$$J = \frac{K_{eqv}^2}{E} = J_e + J_p \tag{5.1}$$

In Eq. 5.1, the parameter $K_{eqv}$ is the equivalent stress intensive factor corrected for plastic deformation in the ligament around the cracked section. The parameter, $J_e$ is the elastic portion of J-integral whereas $J_p$ is the plastic portion of J-integral. The elastic portion $J_e$ can be calculated using Eq. 5.2 which is given as follows:-

$$J_e = 4\varepsilon_0 \sigma_0 a \left(\frac{P}{P_0}\right)^2 F_e \tag{5.2}$$

In Eq. 5.2, $\sigma_0$ and $\varepsilon_0$ is the yield stress and yield strain, respectively. $a$ is the crack depth whereas $P$ and $P_0$ are the applied load and limit load, respectively. $F_e$ is the linear-elastic geometry correction that can be computed using Newman and Raju's geometric function [5]. In fact, this function is one of the most frequently used closed form solution to calculate linear-elastic stress intensity factors of semi-elliptical surface cracks.

Similarly, the plastic component, $J_p$, can be computed using the reference stress approach [6].

$$J_p = \Delta\varepsilon_p \Delta\sigma a \left(\frac{\pi D^2}{4A_{lig}}\right)^{m+1} \frac{\sigma_0}{\alpha' E} F_1 F_e \tag{5.3}$$

In Eq. 5.3, $\Delta\varepsilon_p$ is the plastic strain rate, $D$ is the diameter of the rod, $E$ is the elastic modulus; and $\alpha'$ and $m$ is the hardening parameters. $F_1$ is the geometric correction factors given in [5]. Verification of Eqs. 5.1–5.3 can be found in [7].

Hence, by using Eqs. 5.1–5.3 in addition to the known crack growth data, the crack evolution analysis can be performed using the following methodology.

1. The grain size dependent material properties for the particular grain size are computed using molecular dynamics,
2. The equivalent stress intensity factor $K_{eqv}$ is computed using Eqs. 5.1–5.2,
3. The crack sizes and shapes at the end of each cycle is computed using the experimental result from Bansal [8],
4. Step 2 and step 3 are repeated until the crack growth becomes unstable.

## 5.2 Molecular Simulation

Due to the difficulty in controlling the grain size of nanocrystalline copper (Cu) during material processing as compared to the ease of controlling grain size in simulation, molecular simulation is used to characterize the elastic modulus for different grain sizes in this research.

**Fig. 5.1** A typical
nanocrystalline
microstructure constructed
using Voronoi method

LaMMPS [9], a molecular dynamics code from Sandia National Laboratory, is used. The material response in this research will be modeled using the EAM (embedded-atom method) potential formulated by Folies et al. [10].

In order to investigate the effect of the grain size on the evolution of the crack, Voronoi tessellation methodology [11] is used to construct bulk nanocrystalline Cu with a grain size of 5, 10 and 15 nm, respectively. A cubic simulation cell size of 20 nm by 20 nm by 20 nm is constructed using about 70,000 atoms. Three-dimensional periodic boundary conditions are also enforced throughout this work so as to minimize the surface effect. Figure 5.1 shows a typical nanocrystalline microstructure used in this project.

In order to generate a realistic initial microstructure with no residual stresses, the simulation cell is first subjected to an energy minimization routine using a Polak-Ribière nonlinear conjugate gradient routine at a temperature of 0 K before it is subjected to NPT equilibration at ambient condition (300 K at atmospheric pressure) for 50 ps. The parameter centro-symmetry is also used during post-processing to distinguish the atoms at the grain boundaries from the atoms in the crystal interior. A centro-symmetry value of less than 0.1 is used to define atoms at the grain boundaries.

Deformation of nanocrystalline materials is then performed at a constant strain rate of 1 ps$^{-1}$ in the y-direction using the equations of motion for the ensemble outline in [12] for constant strain rate.

Mechanical stress is computed using the virial theorem [13] whereas the normal strain will be computed as follows:

$$\varepsilon = \ln\left(\frac{L}{L_0}\right) \tag{5.4}$$

In Eq. 5.4, $L$ is the instantaneous length in the y-direction and $L_0$ is its initial length in the y-direction.

**Fig. 5.2** Equi-axial and randomly orientated grain of ECAE copper after subjected to 16 passes [14]

**Table 5.1** Summary of the number of grains and the average grain size for each model

| Number of grains | Grain size computed (nm) | Grain size (nm) |
| --- | --- | --- |
| 4 | 17.6 | 20.0 |
| 4 | 13.0 | 15.0 |
| 17 | 9.3 | 10.0 |
| 123 | 4.96 | 5 |

## 5.2.1 Validation of the Microstructure

Since grain boundary structure at the atomic level of the grain boundary plays a major role in the deformation process, it is critical to model the grain boundaries and the grains as realistically as possible in order of reasonable conclusions from this study.

The texture of Equi-channel angular extrusion (ECAE) Cu, as shown in Fig. 5.2, is that of a random, single-phase poly-crystal with lognormal grain distribution [6]. It is important for our simulation model to have such characteristic.

In order to validate the microstructure used in this research, the average grain size and grain size distribution is computed and compared to the characteristics of the actual microstructures described in the literature [14]. For this benchmark, the nanocrystalline copper with a grain size of 5 nm grain is used. The volume of each grain can be computed by the product of the number of unit cells needed to make up each grain and the volume of each unit cells. Using the computed volume, the grain size can be then computed by assuming these grains to be spherical in shape Table 5.1 summarizes the number of grains and the average grain size for each model.

Even though the error increases as the number of grains in each model decreases, the largest difference is only 13% for 15 nm. Hence, Voronoi methodology is able to model the average grain size accurately.

Even though Fig. 5.3 shows only the grain size distribution for nanocrystalline copper with a 5 nm grain size constructed using the Voronoi Tessellation methodology, it is representative of the other microstructures constructed using

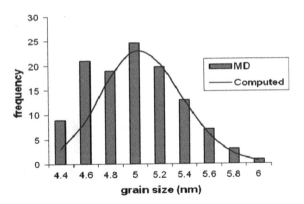

**Fig. 5.3** Grain size distribution for nanocrystalline copper with 5 nm grain size

this methods, Since the grain size distribution of copper has been shown experimentally to follow a log-normal distribution, Fig. 5.3 also compared the estimated grain size distribution from this research with the theoretical log-normal distribution [15, 16].

Figure 5.3 shows that the grain size distribution follows a log-normal distribution with the value of $\mu$ and $\sigma$ computed to be 3.92 and 0.065. The microstructure created in this study appears to represent a realistic grain size distribution. The expected grain size predicted by the log-normal distribution is 5.07 nm. Hence, the Voronoi Tessellation methodology employed in this research can be used to produce microstructure with log-normal distribution and the required average grain size.

As discussed earlier, the texture of ECAE copper is that of a random, single-phase poly-crystal, so there is a need to ensure the randomness of the grain orientation. Unlike the one-dimensional orientations (only [001] tilt boundaries) where the grain disorientation density function is uniform, the disorientation angle of the 3D crystallography is a result of the convolution of two random three-dimensional variables where it is easiest to achieve misorientations near some mean angles. Furthermore, by considering the cubic symmetry of the orientation space, it has been found that Mackenzie distribution function $\rho(\Psi)$ described in Eq. 5.5 [17] could be used to quantify the randomness of the grain disorientation angle.

$$p(\psi) = \frac{2}{15}\left[\left\{3\left(\sqrt{2}-1\right)+\frac{4}{\sqrt{3}}\right\}\sin\psi - 6(1-\cos\psi)\right]$$

$$-8/5\pi\left[2\left(\sqrt{2}-1\right)\cos^{-1}(X\cot\psi/2)+1/\sqrt{3}\cos^{-1}(Y\cot\psi/2)\sin\psi\right]$$

$$+\frac{8}{5\pi\left\{2\cos^{-1}\left\{\frac{(\sqrt{2}+1)X}{\sqrt{2}}\right\}+\cos^{-1}\left\{\frac{(\sqrt{2}+1)Y}{\sqrt{2}}\right\}\right\}}(1-\cos\psi) \qquad (5.5)$$

**Fig. 5.4** Comparison of the grain disorientation density function of the microstructure created in this research with the Mackenzie distribution function

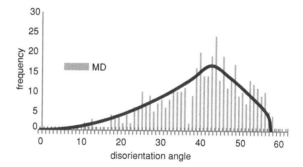

**Table 5.2** Comparison of the fitted Mackenzie parameters of the microstructure created in this research with that of a random texture

| | Mackenzie parameters | |
| --- | --- | --- |
| | Random texture | Simulated microstructure |
| $\bar{\psi}$ | 40.73 | 40.32 |
| $\sigma$ | 11.315 | 10.93 |
| $\Psi$med | 42 | 42 |

In Eq. 5.5, $\Psi$ is the disorientation function and the parameter X and Y are given by Eqs. 5.6 and 5.7.

$$X = \frac{(\sqrt{2} - 1)}{\left[1 - (\sqrt{2} - 1)^2 \cot^2 \frac{\psi}{2}\right]^2} \tag{5.6}$$

$$Y = \frac{(\sqrt{2} - 1)^2}{\left[3 - \cot^2 \frac{\psi}{2}\right]^{0.5}} \tag{5.7}$$

Thus, grain disorientations distribution of the initial structures created in this research can now be benchmarked with the Mackenzie distribution function so as to determine the texture of our microstructure.

Figure 5.4 compares the grain disorientation distribution of the initial structures created in this study with the Mackenzie distribution function. The Mackenzie distribution will only describe a randomly distributed texture for certain value of the mean, $\bar{\psi}$ the standard deviation $\sigma$ and the mean $\Psi$ med.

Since Table 5.2 shows the good agreement between the fitted Mackenzie parameters from our simulated microstructure and that for a randomly distributed texture, the Voronoi Tessellation methodology employed in this study can be used to model randomly textured, single-phase poly-crystalline copper.

**Fig. 5.5** Schematic of the
model used to reproduced
the generalized stacking
fault curve

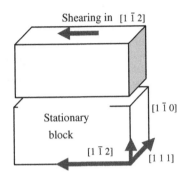

## 5.2.2 Validation of the Inter-Atomic Potential

The inter-atomic potential describes the interaction/movement of atoms in our simulation. Hence, it will be critical to first verify the ability of the inter-atomic potential used in this work to accurately model the deformation of copper. Validation of the potential is achieved through stacking fault energy calculation and modulus computation.

### 5.2.2.1 Stacking Fault Energy

This research repeats the calculation of the generalized stacking fault energy outlined by the work of Zimmerman et al. [18] with the copper EAM interatomic potential. The computed result is then compared with the results for stacking fault energies reported by other researchers to validate the accuracy of the potential used.

A simulation cell with orientation $[1\bar{1}2], [1\,11]$ and $[1\,\bar{1}0]$ in the $x$, $y$ and $z$ direction respectively has been constructed for this run. The size of the simulation cell is $10 \times 40 \times 10$ lattice units on each side. Furthermore, the simulation cell is then sectioned into two blocks by cutting along the {1 1 1} plane as shown in Fig. 5.5.

In order to eliminate the influence of the free surface in the calculation of the stacking fault energy, periodic boundary conditions are enforced in the x and y directions. A uniform shear motion is applied to the top block in the simulation whereas the bottom block is kept stationary. The equation of motion for NVE ensemble is then used to perform the shear calculation at 0 K. The generalized stacking fault energy is then computed by monitoring the energy of each atom near the area where the fault is created. The value of unstable stacking fault energy $\gamma_{usf}$ and the stacking fault energy $\gamma_{sf}$ is calculated as the energy difference between the deformed and undeformed state at the critical point along the generalized stacking fault curve.

Figure 5.6 shows the generalized stacking fault curve of copper from three EAM potential from the literature.

**Fig. 5.6** Generalized
stacking fault curve of copper
from three EAM potential
from the literature review
[12, 19]

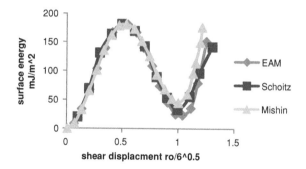

**Fig. 5.7** Computed grain
boundaries volume fraction
as a function of grain size

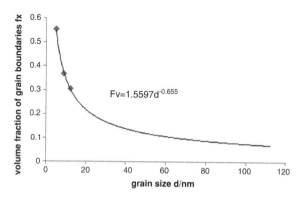

All three inter-atomic potentials show very similar profiles for the generalized
stacking fault energy. The stacking fault energy and unstable stacking fault energy
computed by the copper EAM potentials used in this project are 20 and 180 mJ/m$^2$
respectively which compare very well with the data in Fig. 5.6. Although the EAM
potentials used in this research are not as close to the value of 45 mJ/m$^2$ [20]
predicted by both experimental result and Mishin et al. [21] for copper, this EAM
potential performed pretty well as compared to the other inter-atomic potentials.
Furthermore, the stable and unstable stacking fault energy calculations are consistent with the tight-binding calculation of 162 and 18.2 mJ/m$^2$, respectively.
Hence, the EAM potential used in this research is adequate for accurately modeling the deformation mechanism of copper.

## 5.2.3 Young's Modulus

Due to the difficulty in obtaining the experimental data from the literature at such a
small grain size, the following method is used to validate the modulus computed in
your simulation experiments.

Using a value of less than 0.1 for atoms at the grain boundaries, the volume
fraction of grain boundaries as a function of grain size is computed as shown in
Fig. 5.7.

**Fig. 5.8** The function of
elastic modulus as a function
of grain boundaries volume
fraction

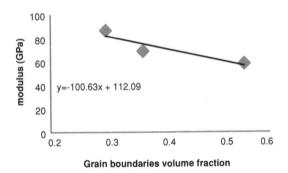

As shown in Fig. 5.7, the expression for volume fraction as a function of grain size is given by

$$f_v = 1.5597 D^{-0.655} \qquad (5.8)$$

In Eq. 5.8, $f_v$ is the volume fraction of the grain boundary atoms whereas $D$ is the average grain size of nanocrystalline Cu. Fig. 5.7 also shows that the volume fraction of grain boundaries increases significantly for grain size smaller than 100 nm (100 nm length scale is often used to formally define nanocrystalline materials). It also emphasizes the importance of grain boundary effects in nano-crystalline copper.

After characterizing the volume fraction of grain boundaries with respect to the grain size, deformation of nanocrystalline materials is then performed at a constant strain rate of 1 ps$^{-1}$ in the y-direction using the equations of motion for the ensemble outline in [18] for constant strain rate. The stress–strain curve can be computed from virial stress and Eq. 5.5 and the elastic modulus is determined from the linear regression analysis of the stress–strain points obtained from the initial portion of the stress–strain curve. Since the rule of mixtures is elected to explain the dependency of the grain size, the modulus will be regressed with a linear relationship.

As shown in Fig. 5.8, the expression for volume fraction as a function of grain size is given by.

$$E = 112.09 + 100.63 f_v \qquad (5.9)$$

Using the rule of mixtures, the elastic modulus for grain boundary free copper is computed to be 112 GPa whereas the modulus of grain boundary modulus is computed to be 12 GPa. The modulus of 112 GPa is consistent with 110 GPa [22] commonly used for microcrystalline copper where the volume fraction of grain boundaries is negligible. Furthermore, the elastic modulus for a 50 nm nano-crystalline copper, computed using Eqs. 5.4 and 5.5, is found to be 100.4 MPa whereas the elastic modulus of a 50 nm ECAE Cu is characterized experimentally to be 100 MPa [7]. This will again validate the accuracy of Eqs. 5.4 and 5.5. Hence, the modulus of the nanocrystalline materials has been accurately charac-terized for this analysis.

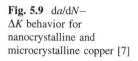

**Fig. 5.9** $da/dN-$ $\Delta K$ behavior for nanocrystalline and microcrystalline copper [7]

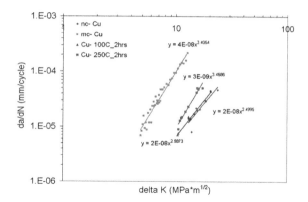

## 5.3 Crack Growth Analysis

Using the material computed in Sect. 5.3 and the methodology proposed in Sect. 5.2 of the paper, crack growth analysis for nanostructure Cu interconnect is performed.

Finite element analysis calculation by Bansal [8] showed that 50 μm pitch microcrystalline copper interconnects experienced a cyclic stress range of 200 MPa and a cyclic plastic strain range of 0.0635 during operation. Hence, this loading condition will be used in this research.

Since nanocrystalline copper will experience only linear-elastic response under these conditions, only linear-elastic portion of J-integral will be needed for the analysis of crack growth in nanocrystalline copper interconnects. However, microcrystalline copper will undergo plastic deformation at this loading condition. Hence, Eqs. 5.1–5.3 must be used for microcrystalline copper. Furthermore, since Ramberg–Osgood hardening equation is capable of modeling inelastic strain response [23], it will be used in this research. The Ramberg–Osgood hardening constants for microcrystalline Cu is obtained from the regression analysis of stress strain data in Bansal [8].

Since the failure morphology of nanocrystalline materials usually consists of dimples several times larger than their grain size [24], the initial crack length is assumed to be four times the average grain size in this research. Using the diameter of copper interconnects to be 25 μm and the crack growth data found in the literature, as shown in Fig. 5.9, the crack evolution of microcrystalline copper interconnect and nanocrystalline copper interconnect subjected to fatigue loading is computed as shown in Fig. 5.10.

Figure 5.9 shows that nanocrystalline copper has a much lower $da/dN-$ $\Delta K$ rate. This is not true at lower $\Delta K$ as shown in Fig. 5.10 with the change over at $\Delta K$ value of 0.463 MPa* $m^{0.5}$.

Even though Fig. 5.10 shows that nanocrystalline copper has a much lower $da/dN-\Delta K$ rate, the crack evolution analysis in Fig. 5.11 shows that the unstable crack growth for microcrystalline copper is found to have occurred at a much

**Fig. 5.10** d*a*/d*N*–
Δ*K* behavior for
nanocrystalline and
microcrystalline copper at
lower Δ*K* values

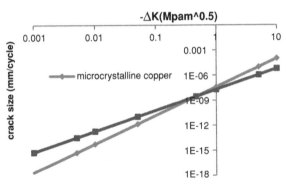

**Fig. 5.11** Crack evolution of
nanocrystalline copper and
microcrystalline copper

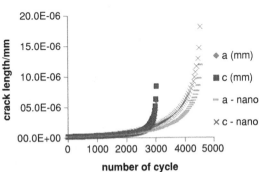

earlier stage as compared to the nanocrystalline copper. Unstable crack growth
occurs at around 3,000 cycles for microcrystalline copper whereas it is expected to
occur at around 4,000 cycles in nanocrystalline copper for this particular loading
condition. This is a 33.3% increase in the crack growth life for nanocrystalline
copper. The reason for this enhanced fatigue life of nanocrystalline copper is due
to its enhanced strength and fatigue crack growth resistance. Since nanocrystalline
copper is still in the elastic regime for this loading condition, less damage will be
induced during each cycle. Thus, nanocrystalline copper is a much better candidate
for the application of interconnects for high pitch applications.

Furthermore, Koh et al. [7] had showed that the total cycle to failure for a 50 μm
pitch nanostructure interconnects subjected to a loading condition of 400 MPa was
2,634 cycles. Hence, the same loading condition is applied to our analysis on
nanocrystalline copper to determine the significance of the crack growth life as
compare to total fatigue life. Microcrystalline copper is not investigated for this
analysis as they will not be able to withstand such a high load. Figure 5.12 shows
the computed crack growth evolution for two aspect ratio of a/c.

Figure 5.12 shows that the unstable crack growth for nanocrystalline copper
occurs at around 600–700 cycles whereas Koh et al. [7] had predicted the total
cycle to failure for nanostructured interconnects to be 2,634 cycles. Hence, the
crack propagation is only one fifth of the total life. This leads to the conclusion that
the long crack growth accounts for a relatively small portion of the total fatigue

**Fig. 5.12** Crack growth evolution for nano-crystalline copper with different initial a/c ratio

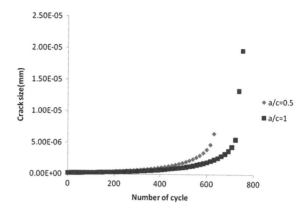

life of the material for the experimental LCF conditions. Hence, initiation of the cracks in the interconnection is the main criterion used to predict its fatigue life.

## 5.4 Conclusions

In conclusion, an accurate closed form solution, for studying 3D semi-elliptical crack growth in round bars subject to uniaxial fatigue loading in both linear-elastic and elastic–plastic low cycle fatigue (LCF) condition, has been developed in this research. Furthermore, molecular dynamics was employed in conjunction to the rule of mixtures to characterize the modulus for nanocrystalline materials as a function of grain size that is needed for the crack growth analysis.

The results indicate that nanocrystalline copper is in fact a suitable candidate for ultra-fine pitch interconnects applications. This study also predicts that crack growth is a relatively small portion of the total fatigue life of interconnects under LCF conditions. Hence, crack initiation life is the main factor in determining the fatigue life of interconnects.

## References

1. http://www.itrs.net/Links/2003ITRS/Home2003.htm
2. Aggarwal O et al (2002) Design and fabrication of high aspect ratio fine pitch interconnects for wafer level packaging. In: Proceedings of 4th Electronic Packaging Technical Conference, Singapore, pp 229–234
3. Chng AC et al (2004) Fatigue life estimation of a Stretched-Solder-Column ultra-fine-pitch wafer level package using the macro-micro modelling approach. In: Proceedings of 54th Electronic Components and Technology Conference, Las Vegas, NV, pp 1586–1591
4. Saxena A (1998) Nonlinear fracture mechanics for engineers. CRC Press, New York
5. Newman JC, and Raju IS Stress-intensity factor equations for cracks. In three-dimensional finite bodies, fracture mechanics: fourteenth symposium—volume I: theory and analysis, ASTM STP 791, Am Soc Test and Mat pp I-238–I-265

6. Findley KO et al (2007) J-integral expressions for semi-elliptical cracks in round bars. Int J Fatigue 29:822–828
7. Koh S et al (2011) Low cycle fatigue crack growth In nanostructure copper. In: Proceedings of EuroSimE 2011, Linz, Austria (to be published)
8. Bansal S (2006) Characterization of nanostructured metals and metal nanowires for chip-to-package interconnections, PhD dissertation. Georgia Institute of Technology
9. Plimpton S (1995) Fast parallel algorithms for short-range molecular dynamics. J Comput Phys 117(1):1–19
10. Foiles SM et al (1986) Embedded-atom-method functions for the FCC metals Cu, Ag, Au, Ni, Pd, Pt, and their alloys. Phys Rev B (Condens Matter) 33(12):7983–7991
11. Du Q et al (1999) Centroidal Voronoi tessellations: applications and algorithms. SIAM Rev 41:637–676
12. Spearot DE (2001) Interface cohesion relations based on molecular dynamics simulations. PhD dissertation. Georgia Institute of Technology
13. Allen MP, Tildesley DJ (1989) Computer simulation of liquids. Oxford University Press, New York
14. Dalla Torre F et al (2004) Microstructures and properties of copper processed by equal channel angular extrusion for 1–16 passes. Acta Mater 52:4819–4832
15. Crow EL, Shimizu K (1988) Lognormal distributions: theory and applications. CRC Press
16. Johnson NL et al (1994) Continuous univariate distributions, vol 1. Wiley, New York
17. Mackenzie JK (1958) Second paper on statistics associated with the random disorientation of cubes. Biometrika 45(1/2):229–240
18. Zimmerman JA et al (2000) Generalized stacking fault energies for embedded atom FCC metals. Model Simul Mater Sci Eng 8(2):103–116
19. Swygenhoven Van et al (2004) Stacking fault energies and slip in nanocrystalline metals. Nat Mater 3(6):399–403
20. Carter CB, Ray ILF (1977) On the stacking-fault energies of copper alloys. Philos Mag 35:189–200
21. Mishin Y et al (2001) Structural stability and lattice defects in copper: ab initio, tight-binding, and embedded-atom calculations. Phys Rev B 63:224101–224106
22. Lu L et al (2000) Superplastic extensibility of nanocrystalline copper at room temperature. Science 287:1463–1466
23. Ramberg W, Osgood WR (1943) Description of stress-strain curves by three parameters national advisory committee for aeronautics
24. Kumar KS, Van Swygenhoven H, Suresh S (2003) Mechanical behavior of nanocrystalline metals and alloys1. Acta Mater 51(19):5743–5774

# Part III
# Molecular Scale Modeling Uses for Carbon Nanotube Behavior

## Introduction

As there is expanding interest in the application of carbon nanotubes (CNT) in many areas of material science, this section deals specifically with aspects of modeling their behavior, specifically the thermal conductivity. In Chap. 6, Fan et al. deal with the thermal conductivity of carbon nanotubes under different stresses including moisture and mechanical. Whereas in Chap. 7, Platek et al. address the effects of specific architectural variables of the carbon nanotube on the thermal conductivity. Both chapters apply molecular dynamics to calculate the thermal conductivity.

(Note: For those unaccustomed to carbon nanotube classification the authors will refer to a specific type of CNT used in their calculations which uses a pair of integers called the chiral vector $(m, n)$. The integer pair represents the number of unit vectors in the graphene sheet used to form the CNT and can be useful to understand roll direction. The type of roll can be understood by looking at the ends of the CNT. For instance, if $m = 0$, the end benzene rings appear to have 1–4 linkages at the CNT terminus and are denoted "armchair"; if $n = m$ the end benzene rings appear to have 1–3 linkages at the CNT ends and are denoted "zigzag". For other indices, the CNTs are denoted "chiral".)

## Part III Chapter List

**Chapter 6:** "Thermal Conductivity of Carbon Nanotube Under External Mechanical Stresses and Moisture by Molecular Dynamics Simulation"
H. Fan, K. Zhang and M. M. F. Yuen

**Chapter 7:** "Influence of Structural Parameters of Carbon Nanotubes on their Thermal Conductivity: Numerical Assessment"
Bartosz Platek, Tomasz Falat and Jan Felba

# Chapter 6
# Thermal Conductivity of Carbon Nanotube Under External Mechanical Stresses and Moisture by Molecular Dynamics Simulation

H. Fan, K. Zhang and M. M. F. Yuen

This paper is based upon "Investigation of Carbon Nanotube Performance under External Mechanical Stresses and Moisture", by H. Fan, K. Zhang, M.M.F. Yuen which appeared in the Proceedings of Eurosime 2007 © Year, IEEE.

**Abstract** In this study, molecular dynamics (MD) simulations were conducted to investigate thermal conductivity of carbon nanotube (CNT) under different conditions. Heat flux was applied on the system and temperature in each region of the single-walled CNT (SWCNT) was calculated from the velocities of atoms from MD simulation. Based on Fourier's law, thermal conductivity was obtained from the heat flux and the calculated temperature gradient along the SWCNT. The MD simulation results showed that the surrounding conditions except for compression stress have a negative effect on the thermal conductivity of CNT.

## 6.1 Introduction

Thermal dissipation is a key issue in electronic packaging design. Ineffective thermal dissipation inside packages can result in not only delamination failure between different layers but also reduce the life cycle of the packages. Compared with traditional thermal interface material (TIM), carbon nanotube (CNT) has excellent mechanical and thermal properties, and is now attracting more attention of researchers as a TIM to dissipate heat from die to heat sink in electronic packaging.

H. Fan (✉) · K. Zhang · M. M. F. Yuen
Department of Mechanical Engineering, Hong Kong University of Science and Technology Clear Water Bay, Kowloon, Hong Kong SAR, China
e-mail: HB.FAN@philips.com

N. Iwamoto et al. (eds.), *Molecular Modeling and Multiscaling Issues for Electronic Material Applications*, DOI: 10.1007/978-1-4614-1728-6_6,
© Springer Science+Business Media, LLC 2012

Investigation of the thermal performance of CNT is rather important for the thermal dissipation in electronic packaging. Due to the technical difficulty of experimental measurement of the CNT thermal conductivity, MD simulation is a powerful technique to investigate heat transport in nanostructure. Molecular modeling represents molecular structures numerically and simulates their behavior with the equations of quantum and classical physics and it is one of the fastest growing fields in science. MD simulation has been successfully applied to predict the thermal conductivity of single-walled CNT (SWCNT) or multi-walled CNT (MWCNT) [1–7]. The predicted thermal conductivity of SWCNT or MWCNT varied from several hundreds to thousands W/km. These MD simulations mainly were focused on the CNTs without considering the surrounding effects. However, CNTs are often subjected to moisture and external stresses, which normally are resulted from CNT-TIM assembly, or CTE mismatch of interface materials connected with the two ends of the CNT during the package working environment. These external stresses can not only result in the deformation of CNT but also change the CNT properties. Guo et al. [8] studied the effect of the tensile loading and electronic field on the mechanical and electrostatic properties. They found that the electronic polarization and mechanical deformation induced by an electric field could result in significant change of the electronic properties of a CNT, and the tensile load changed the energy gap of the tube affecting the field emission properties of CNTs. Ding et al. [9] also found that the electronic conductance depended on the strain of metallic armchair CNTs. However, little attention has been focused on the investigation of external stress effect on the thermal conduction of CNT. The mechanism of thermal conduction of CNT under different environmental conditions is not very clear. How do the CNTs perform under these conditions in real packages? Understanding of these issues is rather important for the design of CNT as TIM in electronic packaging. MD simulation can provide fundamental knowledge to predict these effects on the material properties of CNTs at a fundamental level.

## 6.2 Molecular Dynamics Simulation

Basically, two kinds of MD simulation methods were used for investigation of the heat transfer in CNT. One is the equilibrium molecular dynamics (EMD) method based on Green–Kubo relations [10, 11]. The other is the non-equilibrium molecular dynamics (NEMD) [12] method based on Fourier's law. Due to the difficulty of converging the heat flux and the complexity of the autocorrelation function in EMD method, NEMD method was used to predict thermal conductivity by Fan et al. [6, 7]. The same method was also used in this study.

As we know, the thermal conductivity of SWCNT is dependent on the length, diameter of SWCNT, defects and temperature. In order to avoid other effects on the CNT thermal conductivity, MD simulation was focused on the defect-free SWCNT with a certain length at room temperature in this study. A (10,10)

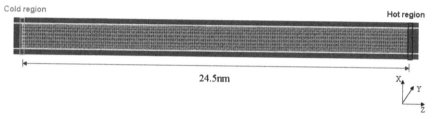

**Fig. 6.1** MD model of SWCNT with a finite length under different conditions

SWCNT model with a length of 24.5 nm was built with the simulation box periodical in $x$ and $y$ directions, as shown in Fig. 6.1. The axial stresses (including tension and compression stresses) or torsion stress, respectively, were applied to the two ends of the CNT. The stress value varied from 0 to 500 Mpa, which are in the range of interfacial stresses occurring in electronic packages.

It is now accepted that open-end CNTs connected with other substrate materials can improve the interfacial adhesion between the CNT and substrates. However, moisture can also penetrate into the tube with larger diameter from the open ends during the assembly. In order to investigate moisture effect on the thermal conductivity of CNTs, MD models with water molecules confined inside the CNT were also built. The water molecules were initially randomly put inside the tube and they were allowed to move freely in all directions. Energy minimization was performed to find the equilibrated structure of the system. The mass ratio of water molecules inside the tube to the CNT varied from 1.75 to 6.50%.

The MD model was divided into N regions along the $z$ axial direction of the CNT. The two regions at the two ends were defined as hot and cold regions respectively. Based on the NEMD algorithm proposed by Ikeshoji and Hafskjold [13], a constant energy was input to the hot region and the same amount of energy was removed from the cold region. The velocities of atoms in both the hot and cold regions were scaled to accomplish the energy transfer from the hot region to the cold region, in which both the energy and momentum are conserved for the whole system. The heat flux in the system was defined as follows:

$$J = \frac{\Delta E}{S \Delta t} \tag{6.1}$$

where $J$ is the heat flux of the CNT, $S$ is the cross-section area of the CNT and $\Delta t$ is the simulation time step.

The instantaneous local temperature in each region can be found by:

$$T_k = \frac{1}{3 n_k k_B} \sum_{i=1}^{n_k} m_i v_i^2 \tag{6.2}$$

where $n_k$ is the amount of atom in region $k$, $k_B$ is Boltzmann's constant, $m_i$ and $v_i$, respectively, are the mass and velocity of atom $i$.

Based on the temperature gradient along the CNT and the heat flux, the thermal conductivity of the CNT, $\lambda$, is obtained by the Fourier law:

$$\lambda = \frac{\langle J \rangle}{\langle \partial T / \partial Z \rangle} \tag{6.3}$$

where $\partial T / \partial Z$ is the temperature gradient along the CNT and the brackets denote a statistical time average.

All the systems were initially equilibrated for 80 ps at room temperature using the ensemble of the constant number of particles, volume and temperature (NVT). A small energy was then imposed on the system. NEMD simulations were conducted based on the algorithm proposed by Ikeshoji and Hafskjold [13] using the ensemble of the constant number of particles, volume and energy (NVE). All the MD simulations run for a long time until a steady-state temperature distribution achieved along the SWCNT. The velocity verlet algorithm was used for integration in all MD simulations. The non-bonded interactions included van der Waals and electrostatic forces. The atom-based summation with a cutoff distance of 0.95 nm was used for the dispersion interactions. All the simulations were performed with an interval of 1 femtosecond (fs) in each MD simulation step. For each kind of MD model, velocities of atoms in each region were averaged over the last 1 ps period to the temperature calculation along the SWCNT.

## 6.3 Results and Discussion

Figure 6.2 showed the temperature profile along the stress-free SWCNT at room temperature. Temperature distributed linearly along the SWCNT and the slope of the fitted straight line showed the temperature gradient. The heat flux was calculated by the energy and the cross-section area with a 3.4 Å thick annular ring. Based on the above equations, the thermal conductivity of the SWCNT under different conditions was calculated for the SWCNT, and the results were listed in Table 6.1. It was found that both the external stresses and moisture had a large effect on the thermal conductivity of SWCNT. The thermal conductivity of CNT with water molecules confined inside the tube was smaller than that of the stress-moisture-free SWCNT and decreased with the increase of moisture inside the tube. The degradation of the thermal conductivity resulted from the increase of the phonon scattering caused by water molecules. Water molecules inside the tube interact with their neighboring carbon atoms in the tube, which act as additional scattering centers resulting in the mean free path of phonon and significantly reduced the thermal conductivity.

From the above MD simulations, it was found that the thermal conductivity of SWCNT was dependent on the applied stress. For those CNT subjected to axial stresses, the thermal conductivity monotonically decreased with the stress changing from the compression to tension. The change of the thermal conductivity

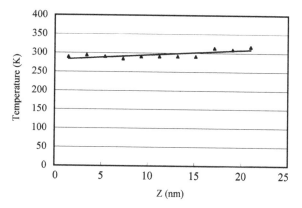

**Fig. 6.2** Temperature distribution along the SWCNT at room temperature

**Table 6.1** Thermal conductivity of CNT under different conditions

| | Free condition | CNT under axial stresses | | | | | | Mass ratio of water molecules to the CNT (%) | | |
| --- | --- | --- | --- | --- | --- | --- | --- | --- | --- | --- |
| | | Compression (Mpa) | | Tension (MPa) | | | | | | |
| | | 150 | 50 | 50 | 150 | 300 | 500 | 1.75 | 3.32 | 6.50 |
| Thermal conductivity (W/mk) | 255.7 | 272.8 | 265.5 | 241.8 | 227.7 | 174.8 | 158.9 | 193.2 | 160.4 | 130.3 |

varied from 4 to 38%. The thermal conductivity of the CNT subjected to torsion stress also decreased with the increase of torsion stress. These results indicated that the stresses had an effect on the transport properties of CNT, not only the electrical conductivity [8, 9] but also the thermal conductivity.

The stress effect on the thermal conductivity can be explained by the kinetic theory. The thermal conductivity of CNT is dominant by phonon scattering and decided by the specific heat capacity, velocity and mean free path of phonon. At the given temperature, the specific heat capacity is the constant, while the velocity and mean free path of phonon are affected by the external stresses applied on the CNT. Picu et al. [14] investigated the stain and size effect on heat transport in nanostructures and found that the lattice thermal conductivity reduced or enhanced by the tensile or compressive load applied on the nanostructure. They addressed that the phonon velocity was proportional to the square root of the stiffness of the nanostructure which increased in compression and decreased in tension. They also pointed out that the variation of mean free path resulted from the lattice anharmonicity was a dominant factor on the strain dependence of the thermal conductivity. The same conclusions were given by Bhowmick and Shenoy [15] who presented a systematic study of strain effect on the thermal conductivity of an insulating solid.

These results are all consistent with the results from MD simulations in this study confirming that the intrinsic thermal conductivity of SWCNT was affected by external stresses. The finding in this study can also be used to explain why the thermal conductivity of a (10,10) CNT is higher than that of a (5,5) CNT. The higher stretching strain along the circumference in a (5,5) CNT made by a rolled graphene sheet reduced the phonon mean free path, which results in lower thermal conductivity, which was also mentioned by Che et al. [1]. However, attention should be paid to the compression stress effect on the thermal conductivity of CNT. Due to the hollow structure of the CNT, CNT under larger compression stress would buckle or even collapse [16], which should heavily affect the phonon transfer and result in degradation of thermal conductivity.

Except for the compression stress, moisture, tensile and torsion stresses have a negative effect on the thermal conductivity of CNT. This will significantly affect the thermal performance of CNT-based assemblies, especially CNT array as TIM in electronic packaging. The induced ineffective thermal dissipation can threaten the reliability of the electronic packages. Therefore, dry environment during the assembly of CNT-based TIM is needed to avoid moisture effect, and CTE matched materials connected with CNT should be considered in the design, which is important to enhance the performance of CNT-based TIM in electronic packaging.

The results from MD simulations predicted the operational environmental effects on the thermal conductivity of SWCNT. The results indicate that more attention has to be paid to the CNT assembly to prevent the environment-induced degradation of the thermal performance of CNT-array TIM in electronic packages. The present MD simulation approach primarily provides a qualitative prediction of the thermal performance of SWCNT under different conditions. Further studies will be focused on the thermal resistance between CNTs and other electronic materials, such as heat sink and silicon chip.

## 6.4 Summary

The study was focused on the operational environmental effects on the thermal conductivity of SWCNT using molecular dynamic simulations. MD results showed that all factors except for compression stress in this study had a negative effect on the thermal conductivity of CNT. The thermal conductivity of SWCNT subjected to axial stress decreased with the stress changing from the compression to tension, while moisture and torsion stress degraded the thermal conductivity of CNT. Therefore, attention should be paid to the CNT assembly and materials selection, which is significant for the thermal dissipation in electronic packaging.

## References

1. Che J, Cagin T, Goddard WA III (2000) Thermal conductivity of carbon nanotube. Nanotechnology 11:65–69
2. Yao Z, Wang J, Li B, Liu G (2005) Thermal conduction of carbon nanotubes using molecular dynamics. Phys Rev B 71:085417

3. Bi B, Chen Y, Yang J, Wang Y, Chen M (2006) Molecular dynamics simulation of thermal conductivity of single-wall carbon nanotube. Phys Lett A 350:150–153
4. Osman MA, Srivastava D (2001) Temperature dependence of the thermal conductivity of single-wall carbon nanotube. Nanotechnology 12:21–24
5. Maruyama S (2003) A molecular dynamics simulation of heat conduction of a finite length single-walled carbon nanotube. Microscale Thermophys Eng 7:41–50
6. Fan HB, Zhang K, Yuen MMF (2006) Effect of defects on thermal performance of carbon nanotube investigated by molecular dynamics simulation. In: Proceedings of the EMAP, Hong Kong, pp 451–454
7. Fan HB, Zhang K, Yuen MMF (2009) The interfacial thermal conductance between a vertical single-wall carbon nanotubes and a silicon substrate. J Appl Phys 106:03430
8. Guo Y, Guo W (2003) Mechanical and electrostatic properties of carbon nanotubes under tensile loading and electric field. J Phys D Appl Phys 805–811
9. Ding JW, Yan XH, Cao JX, Wang DL, Tang Y, Yang QB (2003) Curvature and strain effect on electronic properties of single-wall carbon nanotubes. J Phys Condes Mater 15:439–445
10. Green MS (1954) Markoff random processes and the statistical mechanics of time-dependent phenomena. II. Irreversible processes in fluids. J Chem Phys 22:398–413
11. Kubo R (1957) Statistical–mechanical theory of irreversible processes. I. General theory and simple applications to magnetic and conduction problems. J Phys Soc Jpn 12:570–586
12. Evans DJ, Morriss GP (1990) Statistical mechanics of non- equilibrium liquids. Academic Press, London
13. Ikeshoji T, Hafskjold B (1994) Non-equilibrium molecular dynamics calculation of heat conduction in liquid and through liquid–gas interface. Mol Phys 81:251–261
14. Picu RC, Bocra–Tasciuc T, Pavel MC (2003) Strain and size effect on heat transport in nanostructure. J Appl Phys 93:3535–3539
15. Bhowmick S, Shenoy VB (2006) Effect of strain on the thermal conductivity of solids. J Chem Phys 125:164513
16. Liew KM, Wong CH, He XQ, Tan MJ, Meguid SA (2004) Nanomechanics of single and multiwalled carbon nanotubes. Phys Rev 69:115429

# Chapter 7
# Influence of Structural Parameters of Carbon Nanotubes on their Thermal Conductivity: Numerical Assessment

Bartosz Platek, Tomasz Falat and Jan Felba

**Abstract** This paper focuses on the influence of different parameters of carbon nanotubes (CNT) like length, diameter and chiral vector on its thermal conductivity. To calculate the value of CNT thermal conductivity, a molecular modeling technique was used. For this purpose a special algorithm based on non-equilibrium molecular dynamic was implemented in commercial software.

## 7.1 Motivation

Year in and year out a massive progress in electronic is noticeable. Producers offer faster and more advanced processors, memories, etc. Such integrated circuits have more and more inputs and outputs and their clock frequency is getting higher. It causes more and more power consumption and heat generation by the electronic devices. The heat has to be efficiently dissipated through package structures to a cooler or ambient environment. To improve the heat dissipation, thermal interface materials (TIM) are often used, which decrease the thermal contact resistance between two bodies and make the heat transfer to outer environment more efficient.

TIM composites contain filler particles of materials with high thermal conductivity like beryllium oxide, aluminum nitride, copper, alumina, silver, diamond,

B. Platek (✉) · T. Falat · J. Felba
Faculty of Microsystems Electronics and Photonics,
Wroclaw University of Technology,
ul. Janiszewskiego 11/17, 50-372 Wroclaw, Poland
e-mail: bartosz.platek@pwr.wroc.pl

N. Iwamoto et al. (eds.), *Molecular Modeling and Multiscaling Issues
for Electronic Material Applications*, DOI: 10.1007/978-1-4614-1728-6_7,
© Springer Science+Business Media, LLC 2012

etc. Carbon nanotubes (CNTs) are one of the most interesting and promising materials for fillers. CNT has a unique electrical [1], mechanical [2] and thermal properties. The literature reports that the carbon nanotubes have a very high value of thermal conductivity: 6,600 $Wm^{-1} K^{-1}$ for an individual single-walled CNT [3] and more than 3,000 $Wm^{-1} K^{-1}$ for an individual multi-walled CNT [4].

In the case of a real experiment for the thermal conductivity measurement of such small particles as CNTs, there are many difficulties to solve. A special measuring system has to be prepared with the use of unique techniques (e.g. special substrate with contact electrodes and temperature sensors [5]), practically without the possibility of estimating its measurement accuracy. Moreover, it is very difficult to manipulate a single carbon nanotube. Therefore some methods of orientation changing are used for CNTs, e.g. dielectrophoresis (DEP) [6]. These difficulties motivate the development and use of computer modeling techniques on the molecular scale, e.g. molecular dynamics (MD). In an MD model, atoms in the system have initial positions and e.g. 'random' initial velocity. The masses of atoms are well known, hence the new positions and velocities could be calculated from the Newtonian's equation of motion after a short period of time. This period of time is so-called the time step of simulation ($\Delta t$) and its value is usually in the range of femtoseconds. The whole system is driven by the force field which describes its behavior. The force field is the scalar field of interaction energy in the molecule. It describes the energy as a function of positions of the nuclei in molecule. It allows one to calculate the interatomic chemical bonds and non-bonding potential energy in the system. There are many force fields for different elements available in the literature and implemented in commercial software.

To investigate the thermal conductivity of CNTs by using molecular dynamics simulation, the heat flow through the structure has to be analyzed. There are two ways to impose a heat flux with non-equilibrium molecular dynamics (NEMD). The first one forces heat flow by changing temperature on both ends of structure and then the heat flux $J$ is calculated ($\Delta T \rightarrow J$). This method was described by Hoover and Ashurst in 1975 [7]. The second one was proposed in 1994 by Ikeshoji and Hafskjold [8]. In this way the constant heat flux is imposed and the temperature gradient $\Delta T$ is calculated from temperature distribution along the modeled structure ($J \rightarrow \Delta T$).

## 7.2 Thermal Conductivity in the Stationary State

Heat can be transferred by thermal conduction, convection and radiation. Each of these phenomenons can be stationary or non-stationary. In the stationary case thermal field is only the position function [$T = f(x, y, z)$], while in non-stationary case it is also the time dependent [$T = f(x, y, z, \tau)$] as shown in Fig. 7.1.

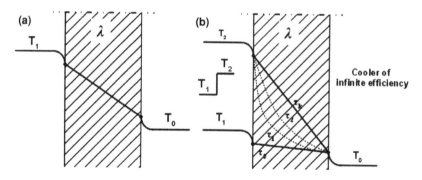

**Fig. 7.1** Temperature distribution in solid body: (**a**) in stationary state, (**b**) in transient state; where there is a stepchange of temperature from $T_1$ to $T_2$ and $\tau$ is the time$(\tau_0 < \tau_1 < \tau_2 < \tau_k)$

In dielectric solid bodies the heat transfer through thermal conductance is the result of crystal lattice vibration. On the other hand, when solid bodies are good electric conductors, heat is transferred by the free electrons (like electric current) which move in a structure of material similar to gas particles. Hence, the heat flow is the result of two phenomena, lattice vibrations (phonons) and movement of free electrons. Because of those two physical phenomena are additive, the total thermal conductivity $\lambda$ is sum of phonon's part $\lambda_p$ and electron's part $\lambda_e$:

$$\lambda = \lambda_p + \lambda_e \tag{7.1}$$

In CNTs the electron density of states is low so the thermal transport via free electrons is negligible. It means that the thermal conductivity is dominated by phonons [8].

In a stationary state the thermal conductivity in one direction is described by Fourier's law (7.2):

$$\lambda = -\frac{J}{\partial T / \partial z} \tag{7.2}$$

where $J$ [Wm$^{-2}$] is the heat flux in $z$ direction, $T$ [K] is a temperature, $z$ [m] is a distance in $z$ direction and $\partial T / \partial z$ [Km$^{-1}$] is a temperature gradient in $z$ direction.

Defining the concept of the mean free path (MFP), each of two types of thermal conductivity discussed above can be divided into two new types: ballistic and diffusive. The mean free path is the average distance of particles between collisions with other particles in an ambient environment. In the ballistic regime the characteristic dimensions are smaller than the MFP. In this regime the thermal conductivity depends on the length of a solid body. For the diffusive type of regime, the length of environment in the movement direction is equal or longer than the MFP, and the thermal conductivity is independent of length, because the heat flux follows with the constant velocity. This velocity is defined by the diffusivity phenomenon.

## 7.3 Molecular Dynamics Simulations

### 7.3.1 Background

To estimate the thermal conductivity of a single-walled CNT (SWCNT) the (NEMD) method (described in [7]) was used. This method provides direct information about transport phenomena because it includes the computation of transport coefficients from flux-force relations being analogous to the macroscopic definition in irreversible thermodynamics. The thermal conductivity was calculated from Eq. 7.2.

Figure 7.2 shows the schematic structure of SWNT used in simulations. Along each CNT the 50 sets of atoms were defined. The group of atoms at one end of nanotube was named as a "Cold" region and on the opposite was named as a "Hot" region. The model was open-ended i.e. without any atoms of hydrogen as a termination. No periodic boundary condition was applied on the model.

In this simulation, constant heat flux was imposed on the system and the resulting temperature gradient was calculated. The advantage of this method lies in the fact that the heat flux which converges slowly is exactly known and need not be calculated. The temperature and its gradient converge rapidly in the local regions. They were calculated over time and over particles in each region of CNT as well. The heat flux was applied by adding a fixed quantity of kinetic energy $\Delta E_\kappa$ in the "Hot" region and subtracting the same amount of energy in the "Cold" region in each iteration. In this method, the heat flux is added into (or extracted from) the "Hot" (or "Cold") region by rescaling the velocity in the region in following way:

$$v_i' = R \cdot v_i - v_{\text{sub}} \tag{7.3}$$

$$v_{\text{sub}} = \frac{(R-1)\sum m_i \vec{v}_i}{\sum m_i} = \frac{(R-1)\vec{P}}{N \cdot m_i} \tag{7.4}$$

where $v_i$ and $v_i'$ are the velocities of atom with mass $m_i$ before and after rescaling, respectively; $N$ is number of atoms, $R$ is a scaling factor and small velocity $v_{\text{sub}}$ maintains the momentum $P$ before and after the energy addition at zero so that the reservoirs remain stationary, and the summation is over all the particles in each region. Energy conservation requires that the added energy $\Delta E_\kappa$ (or extracted energy, by setting $\Delta E_\kappa$ negative) is:

$$\Delta E_\kappa = \frac{1}{2}\sum_{i=1}^{N} m_i(v_i^2 - v_i'^2) \tag{7.5}$$

**Fig. 7.2** Schematic structure of CNT used in NEMD simulation; $k$ corresponds to one of the 48 internal sets of CNT

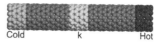

Cold         k         Hot

Equations 7.3–7.5 determine the scaling factor for given $\Delta E_\kappa$ as:

$$R = 1 + \frac{-B + \sqrt{B^2 + 4 \cdot \Delta E_\kappa}}{2 \cdot A} \tag{7.6}$$

where $A$ and $B$ are temporary factors used to velocity scaling which are calculated as follows:

$$A = E_\kappa + \frac{1}{2}\left(\frac{P}{N}\right)^2 \sum_{i-1}^{N} \frac{1}{m_i} - \frac{\vec{P}}{N} \sum_{i=1}^{N} \vec{v}_i \tag{7.7}$$

$$B = 2 \cdot E_\kappa - \left(\frac{\vec{P}}{N}\right) \sum_{i=1}^{N} \vec{v}_i \tag{7.8}$$

The temperature in the $k$-region is obtained by the average kinetic energy of atoms in this region divided by the degrees of freedom and the Boltzmann constant $k_B$:

$$T_\kappa = \frac{1}{3 \cdot N_\kappa \cdot \kappa_B} \sum_{i=1}^{N} m_i v_i^2 \tag{7.9}$$

With the constant energy $\Delta E_\kappa$ added into (or extracted from) the "Hot" (or "Cold") region, heat flux can be calculated:

$$J = \frac{\Delta E_\kappa}{S \cdot \Delta t} \tag{7.10}$$

where $S$ is the cross-sectional area of analyzed SWNT and $\Delta t$ is the time step used to conduct velocity rescaling (e.g. 1 fs). The cross-sectional area is the annulus with the difference between outer and inner radius equal to 3.4 Å-i.e. the thickness of single graphene layer (doubled van der Waals radius of carbon) [9]. The thermal conductivity of analyzed structure is directly calculated from Fourier's law (Eq. 7.2).

### 7.3.2 The Simulation Procedure

Non-equilibrium molecular dynamics was implemented in the materials studio software. At the beginning, the CNT was created in material studio visualizer. NEMD simulations were then performed by using the Discover module with the

**Fig. 7.3** The exemplary
algorithm of the simulation
procedure

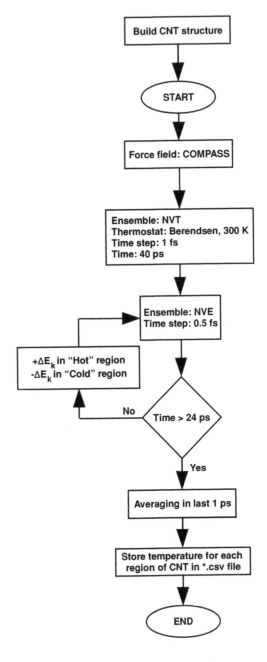

exemplary algorithm of simulation as presented in Fig. 7.3. The algorithm was accomplished by using BTCL language, which is an extension of the tool command language (TCL) which has additional commands that facilitate the manipulation with objects.

The condensed-phase optimized molecular potentials for atomic simulations studies (COMPASS) force field was used to calculate interatomic chemical bonds and non-bonding potential energy. It was followed by the canonical ensemble of the constant number of particles, volume and temperature (NVT) with Berendsen's thermostat for a few tens ps. It makes the system thermodynamically stable at desired temperature (e.g. 300 K) before conducting further calculations.

NEMD simulation was then conducted using the micro-canonical ensemble of constant number of particles, constant volume and constant energy NVE within a specified time (tens or hundreds of picoseconds). During this time the uniform heat flux were imposed on the system and the constant temperature gradient along the CNT was obtained by Eq. 7.9. Velocities of atoms in each region were averaged over some period of time of final stage of simulation to calculate temperature gradient along the CNT.

### 7.3.3  The Impact of Carbon Nanotube's Length on its Thermal Conductivity

The effect of CNT length on its thermal conductivity was done by using SWCNT (10, 10) with different lengths varying from 50 to 400 nm. Each SWCNT was divided into 50 equal regions with "Hot" and "Cold" on both ends as shown in Fig. 7.2.

NEMD simulations were conducted using NVT ensemble at 300 K for 40 ps with 1 fs time step. It was followed by NVE simulation for 24 ps with 0.5 fs time step, in which the kinetic energy in "Hot" and "Cold" regions was changed in every time step. During the simulation, the kinetic energy was $\Delta E_\kappa = 5.18 \times 10^{-22}$ J. This value of energy corresponds to temperature changes by about 30 K in "Hot" and "Cold" regions. The temperature along the CNT was calculated and averaged over the last 1 ps period. For each length of CNT, the simulation was performed five times and average value and corresponding standard deviation were determined, as shown in Fig. 7.4.

Transport of phonons in solids depends on scattering process on lattice vibration. When structure sizes are smaller than MFP, the movement of phonons transfers without dissipation. The research shows that the MFP calculated from measurement for SWCNT (10, 10) at 300 K is even 750 nm [10]. It can explain why the thermal conductivity rapidly rises with increasing length of the CNT in the range 50–400 nm.

### 7.3.4  The Impact of Carbon Nanotube's Diameter on its Thermal Conductivity

To investigate effect of the diameter of CNTs on their thermal conductivity, the 14 models of zig-zag nanotubes were built with diameters varying from 0.548 nm for

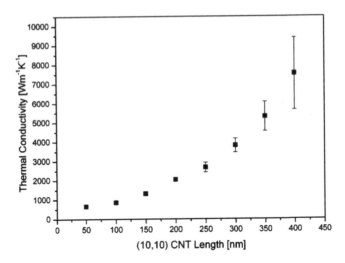

**Fig. 7.4** The influence of (10, 10) SWCNT length on its thermal conductivity

chiral vector (7, 0) to 1.566 nm for chiral vector (20, 0). All structures had the same length of 213 nm.

Using the same procedure as the previous section, NEMD simulations were conducted using NVT and NVE ensembles. The kinetic energy $\Delta E_\kappa = 5.18 \times 10^{-22}$ J in "Hot" and "Cold" regions was exchanged in every time step. The temperature along the CNT was calculated and averaged over last 1 ps period. For each length of CNT the simulation was done four times to determinate the mean value and corresponding standard deviation of the results as shown in Fig. 7.5.

The thermal conductivity for each CNT is around 2,000 $Wm^{-1}$ $K^{-1}$ with a common range of the standard deviation, so it can be concluded that the diameter of zig-zag single-walled carbon nanotube has no effect on their phonon part of thermal conductivity.

## 7.3.5 The Impact of Carbon Nanotube's Chiral Vector on its Thermal Conductivity

The effect of CNT chirality on its thermal conductivity was investigated by using 11 SWNT models with different chirality. The chirality of each CNT was set by changing a chiral vector $(m, n)$ i.e. the way the graphene sheet is wrapped. In the simulation, $m$ was set to 10 and $n$ varied from 0 (zig-zag) to 10 (armchair), and the diameters of CNT varies from 7.83 nm for zig-zag to 13.56 nm for armchair CNT. The simulations were performed with lengths of 43 and 100 nm.

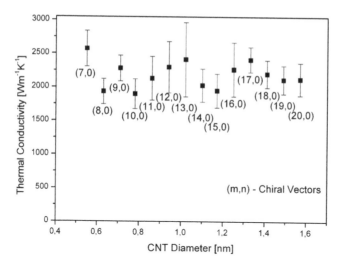

**Fig. 7.5** The influence of zig-zag SWCNT on its thermal conductivity

The NVE ensemble simulation was conducted for 20 ps at 300 K with 1 fs time step for both short CNTs (43 nm) and long CNTs (100 nm). It was followed by NVE simulation. The value of kinetic energy changed in "Hot" and "Cold" regions for short CNT was $\Delta E_K = 1.03 \times 10^{-21}$ J and for long $\Delta E_K = 5.18 \times 10^{-22}$ J. The time of simulation was set on 120 ps with 1 fs time step for 43 nm CNT and 24 ps with 0.5 fs time step for 100 nm CNT. The temperature distribution used for calculations of thermal conductivity was obtained by averaging of temperature in each region over last 1 ps period of simulation for short CNTs and in last 2 ps period for long CNTs. For each type of CNT the simulations were conducted to obtain the standard deviation of results as shown in Figs. 7.6 and 7.7.

The calculated average thermal conductivities varied in the range 206–367 $Wm^{-1} K^{-1}$ for different CNTs. The results for long CNTs were not as scattered as those for short CNTs and the statistical dispersion of results were lower than those for short CNTs.

## 7.3.6 Verification and Model Validation: Thermal Conductivity of Silicon Nanocluster

The idea of model validation and verification of used test procedure is based on the calculation and the conductivity of nanocluster of material with known thermal properties. In the present study, nanoclusters made of silicon were analyzed. The bulk value of silicon thermal conductivity at room temperature is estimated to about 124 $Wm^{-1} K^{-1}$ [11].

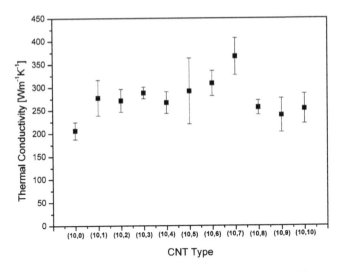

**Fig. 7.6** The thermal conductivity for different chirality of 43 nm long CNTs

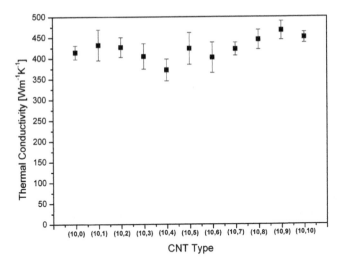

**Fig. 7.7** The thermal conductivity for different chirality of 100 nm long CNTs

The thermal conductivity of two silicon nanoclusters in the form of cuboid with cross-sectional area about $1 \times 1$ nm and different length (300 and 500 nm) was analyzed.

The whole simulation procedure was similar to those used for CNTs. Just as in the case of CNTs, the nanoclusters were divided into 50 equal sections (i.e. "Hot" and "Cold" on both ends and 48 intrinsic sets). However, the simulation parameters have been modified and adopted to silicon nanoclusters. Initially the NVT ensemble was used for both nanoclusters and they were thermodynamically equilibrated at 300 K by making use of Berendsen's thermostat within 32 ps with 1 fs time step. In the next step (micro-canonical NVE ensemble) the kinetic energy $\Delta E_\kappa$ equal $5.18 \times 10^{-23}$ J was applied and the atom velocity rescaling was performed during 1,800 ps. Time step $\Delta t$ was set on 6 fs for 300 nm long nanocluster and on 4 fs for 500 nm long nanocluster. The temperature gradients along the nanoclusters were calculated from the velocity of atoms in each sub-region after averaging over last 1 ps period in case of 300 nm nanocluster and over last 4 ps in case of 500 nm nanocluster.

Based on the obtained simulation results, the thermal conductivity of both nanoclusters were calculated. The values of thermal conductivity were 59.0 $Wm^{-1} K^{-1}$ for 300 nm long nanocluster and 73.1 $Wm^{-1} K^{-1}$ for 500 nm long nanocluster.

The lower thermal conductivity than expected value may be due to the fact that the length of the simulated objects was relatively short compared with the phonon MFP. The MFP at room temperature is estimated for silicon to be around 300 nm [12, 13] (250 nm [14]); for SWNT around 750 nm [15]; and for graphene around 775 nm [16].

Nevertheless, calculated values of thermal conductivity of silicon nanoclusters are comparable with those presented in the literatures. For example, measured out-of-plane thermal conductivity of thin (<1 μm) monocrystalline silicon layers at room temperatures is up to 50% smaller than the bulk value (below 100 $Wm^{-1} K^{-1}$) [12, 17]. Therefore, it can be concluded that used algorithm and simulation procedure is correct and valid, but the further study with longer nano objects is required to get more accurate results within the diffusive regime.

## 7.4 Summary

The non-equilibrium molecular dynamics for calculations of carbon nanotubes' thermal conductivity was presented. In this case molecular dynamics simulations are very useful to predict physical, mechanical, chemical or thermal properties of such small objects as carbon nanotube because of difficulties in manipulating the real structure to perform a test.

Using this technique, effects of carbon nanotube's structural parameters on thermal conductivity were investigated. Simulation results show that the length of CNT has the largest effect on thermal conductivity. While, the diameter and chiral vector have insignificant impact on thermal conductivity.

The simulations of silicon nanocluster further demonstrated property the described methodology of computations.

**Acknowledgments** This work was performed as part of the framework for the "CArbon NanOtubes/ePoxY composites (CANOPY)" project; Eurypides contract no. 06-176. Authors acknowledge Wroclaw Center for Networking and Supercomputing (WCSS) for the possibility of using modeling software and hardware.

# References

1. Saito R et al (1992) Electronic structure of chiral graphene tubules. App Phys Lett 60:2205
2. Ruoff RS et al (2003) Mechanical properties of carbon nanotubes: theoretical predictions and experimental measurements. C R Physique 4:993–1008
3. Berber S et al (2000) Unusually high thermal conductivity of carbon nanotubes. Phys Rev Lett 84(20):4613–4616
4. Philip K et al (2002) Mesoscopic thermal transport and energy dissipation in carbon nanotubes. Physica B 323:67–70
5. Yang W et al (2008) Mounting multi-walled carbon nanotubes on probes by dielectro-phoresis, diamond and related materials. Diam Relat Mater 17:1877–1880
6. Ashurts WT, Hoover WG (1975) Dense-fluid shear viscosity via nonequilibrium molecular dynamics. Phys Rev A 11:658
7. Ikeshoji T, Hafskjold B (1994) Non-equilibrium molecular dynamics of heat conduction in liquid and through liquid-gas interface. Mol Phys 81:251–261
8. Fischer JE (2005) Carbon nanotubes: structure and proprties. In: Gogotsi Y (ed) Nanotubes and nanofibers. CRC Press, Boca Raton, pp 1–36
9. Bondi A (1964) Van der Waals volumes and radii. Phys Chem 68(3):441–451
10. Yu C, Shi L, Yao Z, Li D, Majumdar A (2005) Thermal conductance and thermopower of an individual single-wall carbon nanotube. Nano Lett 5:1842–1846
11. Material property database: http://www.matweb.com
12. Ju YS, Goodson KE (1999) Phonon scattering in silicon films with thickness of order 100 nm. Appl Phys Lett 74:3005–3007
13. Pascual-Gutiérrez JA, Murthy JY, Viskanta R (2009) Thermal conductivity and phonon transport properties of silicon using perturbation theory and the environment-dependent interatomic potential. J Appl Phys 106:063532
14. Ju YS (2005) Phonon heat transport in silicon nanostructures. Appl Phys Lett 87:153106
15. Hepplestone SP, Srivastava GP (2007) Lowtemperature mean-free path of phonons in carbon nanotubes. J Phys Conf Ser 92:012–076
16. Ghosh S et al (2008) Extremely high thermal conductivity of graphene: prospects for thermal management applications in nanoelectronic circuits. Appl Phys Lett 92:151911
17. Asheghi M et al (1997) Phonon-boundary scattering in thin silicon layer. Appl Phys Lett 71:1798–1800

# Part IV
# Molecular Methods to Understand
# Mechanical and Physical Properties

## Introduction

Part IV deals with molecular modeling methods used in physical property determination. The materials used in the three chapters represent diverse material interfaces ranging from organosilicates to epoxies as well as self-assembled monolayers as copper-epoxy coupling agents. Chapter 8 (Yuan et al.) discusses methods to obtain modulus and density properties for organosilicates, specifically porous silicate interfaces which are of importance in low-k dielectrics, and interestingly Yuan describes the Si site structure variation on the properties. The methods used for modeling the properties were derived from models on carbon nanotube buckling, which are also described, making this chapter a nice topical transition between Section III and Section IV. Chapter 9 (Wong et al.) discusses coupling agent work for an epoxy-copper interface, specifically modeling the effectiveness of the Cu-alkylthiol unit as the coupling agent. Molecular models qualitatively showed the effectiveness, tested both with and without moisture and compared to experiment. Chapter 10 (Hölck et al.) demonstrates the use of molecular modeling in the determination of a variety of polymer properties specifically targeted for epoxy molding compounds. In this chapter, both the modeling and the experiments used to validate the models are shown. Significantly, methods to correctly depict cross-linked epoxies are described, and from these structures adequate simulations of density, diffusion and bulk properties are derived. In addition epoxy-silicate interface models are shown including the work of adhesion.

# Part IV Chapter List

**Chapter 8:** "The Mechanical Properties Modeling of Nano-Scale Materials by Molecular Dynamics"
C. Yuan, W. D. van Driel, R. Poelma, G. Q. (Kouchi) Zhang

**Chapter 9:** "Molecular Design of Self Assembled Monolayer (SAM) Coupling Agent for Reliable Interfaces by Molecular Dynamics Simulation"
C. K. Y. Wong, H. Fan, G. Q. (Kouchi) Zhang, M. M. F. Yuen

**Chapter 10:** "Microelectronics Packaging Materials: Correlating Structure and Property Using Molecular Dynamics Simulations"
Ole Hölck, Bernhard Wunderle

# Chapter 8
# The Mechanical Properties Modeling of Nano-Scale Materials by Molecular Dynamics

C. Yuan, W. D. van Driel, R. Poelma and G. Q. (Kouchi) Zhang

This paper is based upon "The chemical-mechanical relationship of the SiOC(H) dielectric film", by Cadmus Yuan, O. van der Sluis, G.Q. Zhang, L.J. Ernst, W.D. van Driel, R.B.R. van Silfhout and B.J. Thijsse which appeared in the Proceedings of Eurosime 2007 © Year, IEEE, and "The mechanical influence of the porosity and nano-scale pore size effect of the SiOC(H) dielectric film", by Cadmus Yuan, Amy E. Flower, Olaf van der Sluis, G.Q. (Kouchi) Zhang, Leo J. Ernst, Mohammed Cherkaoui, Willem D. van Driel which appeared in the Proceedings of Eurosime 2008 © Year, IEEE.

**Abstract** We propose a molecular modeling strategy which is capable of modeling the mechanical properties on nano-scale low-dielectric (low-$k$) materials. Such modeling strategy has been also validated by the bulking force of carbon nano tube (CNT). This modeling framework consists of model generation method, boundary condition setting and result extraction method. For the amorphous silicon-based low-dielectric (low-$k$) material, the impact of the porosity and pore size upon the elastic modulus are modeled. Due to the electronic requirement of advanced electronic devices, low-$k$ materials are in demand for the IC backend structure. However, due to the amorphous nature and porosity of this material, it exhibits low mechanical stiffness and low interfacial strength, as well as inducing numerous reliability issues. The mechanical impact of the nano-scaled pore,

C. Yuan (✉) · G. Q. (Kouchi) Zhang
Delft Institute of Microsystems and Nanoelectronics (Dimes),
Delft University of Technology, Mekelweg 6, 2628 CD, Delft, The Netherlands
e-mail: cadmus.yuan@tno.nl; c.a.yuan@tudelft.nl

C. Yuan
Material Technologies, Netherlands Organisation for Applied Scientific Research (TNO), De Rondom 1, 5612 AP, Eindhoven, The Netherlands

W. D. van Driel · R. Poelma
Department of Precision and Microsystem Engineering,
Delft University of Technology, Delft, The Netherlands

N. Iwamoto et al. (eds.), *Molecular Modeling and Multiscaling Issues for Electronic Material Applications*, DOI: 10.1007/978-1-4614-1728-6_8,
© Springer Science+Business Media, LLC 2012

including the porosity ratio and pore size, is simulated using molecular dynamics on the mechanical stiffness and interfacial strength. A fitting function is formulated based on the continuum homogenous theory and atomic interaction in nano-scale. The simulation results are fitted into analytical equation based on the homogenous theory.

## 8.1 Introduction

As feature sizes for advanced ICs continue to shrink [1, 2], the semiconductor industry is focusing the technology to minimize the intrinsic time delay for signal propagation, quantified by the resistance-capacitance (RC) delay. The increasing demands for the electronic performance of the IC wiring have recently driven the replacement of aluminum trace with copper trace, and alternative materials for $SiO_2$ film having a lower dielectric constant [3]. These alternative low-$k$ materials are silsesquioxane-based material, silica-based material, organic polymers and amorphous carbon. However, the siliconoxide-based low-$k$ materials, such as OSG, are preferred in the industry because the fabricating processes of these materials exhibits higher IC process compatibility and yielding rate. In the silica-based matrix material, the $k$ value can potentially be reduced via two routes. First, chemically, one can replace oxygen by carbon, hydrogen (organosilicate glass, OSG), or fluorine (fluorinated silica glass, FSG). Second, one can physically generate porosity within the silica-based low-$k$ material. The softness of the low-$k$ material introduces enormous proess, mechanical and reliability issues in the IC backend. And low-$k$ material is one of the key issues/challenges in the development of the advanced IC interconnect [1].

The silicon oxide based low-$k$ materials SiOC(H), also called black diamond, illustrated in Fig. 8.1 are preferred by industry because the fabricating processes of this materials exhibits high IC process compatibility and high yielding rate. The k value can be reduced in two ways: either chemically by replacing oxygen by the methyl groups or H, OH or physically by generating porosity within the material. The different Si atoms are indicated with the usual denomination related to the number of O atoms linked to them: mono (M), di (D), tri (T) and quadri (Q)–functional group. The remaining links are of the type Si-R, where R is the –$CH_3$, O and OH functional group [4]. In addition, when functional group is replaced by a silanol group, it is indicated with OH as superscript. Figure 8.1b illustrates the groups of Q, T, D and M.

Among the materials of advanced IC backend structures, the low-$k$ material has low mechanical stiffness, approximately 5–15 GPa. Experiments [4] show that enhancing the Young's modulus of the low-$k$ material will increase the interfacial toughness of SiOC(H)/TaN interface, which is known as the most critical interface in these structures. Among all the enhancement methods, the ultraviolet (UV) curing is preferred because the SiOC(H) film can perform the enhancement of the

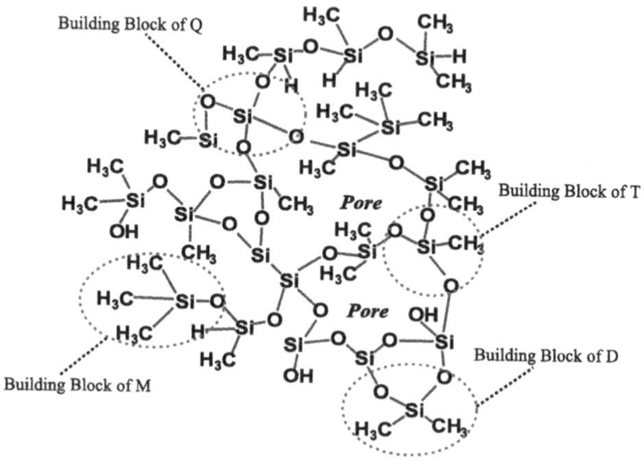

**Fig. 8.1** Atomic structure of BD material

mechanical strength without much loss of the dielectric characteristic. However, the relationship between the chemical composition, porosity and mechanical properties remained unclear, and a trial-and-error design method is still common practice in the design/fabrication of the low-$k$ material in the industry. Therefore, in this study, an atomic modeling method is developed, which is capable to analyze amorphous silica-based material with porosity, to systematically study relation between mechanical characteristics of the SiOC(H) low-$k$ film and it's chemical structure.

Theoretically, the amorphous nature of the SiOC(H) film together with the porosity increases the difficulty to directly simulate its nano-scaled mechanical response. Due to the amorphous nature, the atomic structure can be hardly defined. The void in the SiOC(H) molecule occurs randomly, and the size of the void should be also carefully considered. According to the literature, the complicate molecule (like SiOC(H) film) can be modeled when the accurate atomic structures and the potential functions are available. Falk and Langer [5] have applied the 12-6 Lennard–Jones potential function to describe viscoplastic deformation in amorphous solids.

In our previous research, the molecular modeling is applied to simulate the mechanical response of the low-$k$ material [6–9]. The simulation is comprised of two procedures.

A reasonable topology and stereochemical structure of the amorphous low-$k$ material is generated, followed by a calculation of the mechanical stiffness by molecular dynamics (MD). The atomic topologies of the low-$k$ film and its bonds are generated based on the experimental chemical composition. Through the MD simulation, the Young's modulus is extracted. The trends of the simulated results are validated by the experimental results.

Based on the molecular model from previous step, the interface model, which consists of amorphous low-$k$ material and silica is generated. To study interface

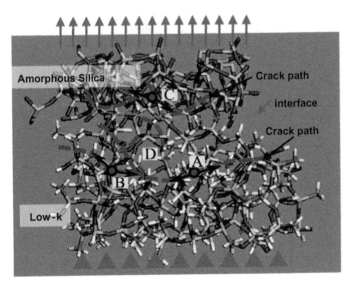

**Fig. 8.2** Illustrates the crack propagation (*upper and lower dashed lines*) through the interface (*middle dashed lines*), where the boundary/loading conditions are also shown

strength depend on the porosity [7] and pore at/near interface (Fig. 8.2), respectively.

In this paper, we will present the molecular modeling strategy, including the 2 steps described above and the result management techniques to approach the experimental validation.

## 8.2 Molecular Modeling for Nanoscale Material

An atomistic method is established herein to predict the mechanical stiffness parameter, which is represented by the Young's modulus, of the nano-scaled structure. The nano-scaled specimens are simulated by the MD method with an additional energy minimization procedure.

A bar model is established as illustrated in Fig. 8.3, where one end of the bar is fixed and the opposite end is applied a displacement. The applied displacements and reaction forces which obtained at the fixed end are used to extract the Young's modulus by the elasticity theory. Due to the small deformation assumption of elasticity [10], the total amount of the longitudinal deformation should be less than 1.0% of the total length of the specimen. Moreover, based on Saint-Venant's principle [10], a model with high aspect ratio ($L/h$) is required to prevent boundary effects, as illustrated in Fig. 8.3. The loading and boundary conditions are applied at the longitudinal direction. Moreover, due to the linearity assumptions, reaction force outputs are linear with the externally applied displacement. The reaction forces $\vec{F}^i$ ($i$ represent the $i$th substep) at the fixed end can be extracted either by the

**Fig. 8.3** Illustration to bar loading model

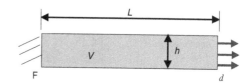

**Fig. 8.4** Illustration scheme of the constrained atoms

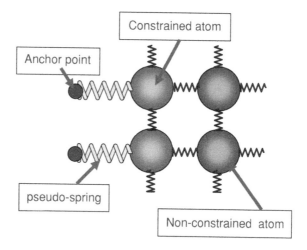

force of the pseudo-spring of the anchor point (illustrated in Fig. 8.4) or the energy gradient of the fixed atoms.

According to linear elasticity theory, the mechanical deformation of the uniaxially loaded bar can be represent as: $\Delta d = FL/EA$ [10], where $F$, $E$, $L$ and $A$ represent external mechanical force, Young's modulus, initial length and initial cross section area of the specimen, respectively.

This modeling setup and boundary condition has been validated by the bulking of the carbon nano tube (CNT): A single-wall CNT with the applied boundary conditions is shown in Fig. 8.5. The ends of the CNT are fixed to linear elastic tethers (springs). The tethers have a specified stiffness and reference length. The tethers are fixed between the atoms at the ends of the CNT and the anchor points. The anchor points are fixed points in a Cartesian reference frame. When the anchor points at one boundary are given a small incremental displacement, the tethers are stretched and a restoring force is created within the tethers.

The following CNT reference configuration was used for the validation of the simulation results. The model consists of a zig-zag, fixed–fixed, single-wall CNT. Zigzag refers to the chirality of the CNT, which is (n;m) = (10; 0) [11]. The CNT geometrical dimensions are: $L = 5.26$ nm the CNT length and $D = 0.78$ nm the CNT diameter.

The simulations are performed at a room temperature of 300 K. Figure 8.6 shows the load displacement curve of the reference CNT under axial compression.

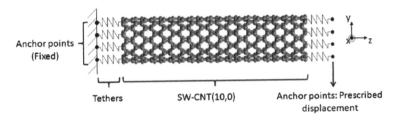

**Fig. 8.5** Schematic illustration of a single-wall (SW) CNT and the approach used for prescribing the boundary conditions and extracting the reaction forces during the simulations

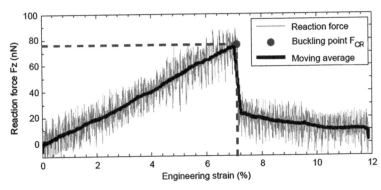

**Fig. 8.6** The load carrying capability of a fixed–fixed CNT at room temperature, before and after buckling. The noise on the reaction forces is caused by the vibration of the atoms at room temperature

From 0 to 7.2% strain, the required compressive force increased linearly until it peaked at 76 nN and the structure buckles. After buckling the load carrying capability is considerably lower. The simulation results are in good agreement with the results (60 nN and a critical strain of 6.7%) of a similar sized doubly clamped CNT obtained from literature [12]. The post buckling behavior also agrees well with the simulation results of [12], considered that different forcefields and simulation tools were used.

## 8.3 Atomistic Model of SiOC(H)

Due to the amorphous and porous nature of the low-*k* material, a unique modeling technique is used to introduce the atomic structure into the MD software. Basically, the chemical structure of the low-*k* material is classified as four different basic building blocks, which has 4, 3 and 2 connection capabilities. These building

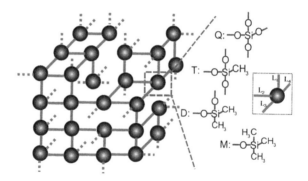

**Fig. 8.7** Generating algorithm of SiOC:H: the pre-defined framework. The nodes will be replaced by either Q, T, D, M or pore

blocks are distributed into a pre-defined framework, where the ration between building blocks follows the experimental result.

A pre-defined framework, which is made up of $SiO_2$ tetrahedral sharing corners, is established. The nodes (gray spheres) and links (red lines) are defined as the possible location of basic building blocks and the possible links (Si–O–Si bonds) between these blocks, respectively (Fig. 8.7). Each node, in this framework, has four links connected to it. The geometrical distance between two nodes is approximately defined as 0.3 nm. Instead of a diamond-like framework, a cubic grid is chosen to provide more free space for the structural relaxation. The building blocks are sequentially distributed into each node.

The distribution obeys the following rules:

(a) Chemical nature of building blocks: When the pore, Q, T, D or M is distributed onto the node, the total link of that node is fixed to zero, four, three, two and one, respectively.
(b) Average distribution: The local high concentration of specific type of building blocks should be prevented. Hence, a reliable random number generator is used to obtain the uniform distribution.
(c) Minimal numbers of dangling bonds: Because the dangling bonds are not physically favored, reducing them will easily lead to a minimal potential energy state.

For the generating of the pore in low-$k$ model under the control of the volume fraction of the pore, the trick of multiple pore generating algorithm is used. As illustrated in Fig. 8.8, additional pore(s) is placed next to the existing pore(s) where has been located in the generating algorithm shown previously. Moreover, the numbers of the void (as illustrated in Fig. 8.8), is controlled. Therefore the volume fraction of the porosity is controlled.

After the molecular topology is obtained as the previous steps, the minimization procedure of total potential energy is applied to obtain the most likely stereo chemical structure of SiOC:H. A non-optimized molecular model might exhibit a non-zero reaction force under zero external deformation applied to the system, due to high intrinsic expanding/shrinking stress of the molecule. The minimization step comprises a continuous iteration of the local perturbation of the atom coordinate

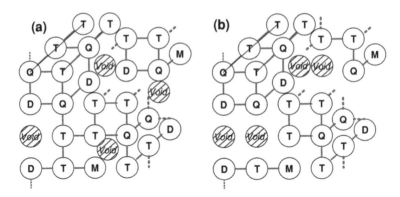

**Fig. 8.8** Illustration of the varying the pore size. **a** Unit pore size, **b** double pore size

and global perturbation of the system. The global perturbation is necessary especially for achieving the stress-free status of large SiOC:H molecule.

## 8.4 MD Simulation Results for Mechanical Properties

In order to verify the accuracy of the proposed method, two SiOC(H) models, named A1 (shown in Fig. 8.9) and A2, having similar chemical composition as the SiOC(H) before and after UV treatment, have been generated. The properties of model A1 and A2 are listed in Table 8.1. It demonstrates the side view and cross sectional view of model A1, where the dark yellow, red, gray and white spheres represent, respectively, the silicon, oxygen, carbon and hydrogen atom. In order to prevent boundary effects, the length and cross section size of both cases are chosen as approximately 10 nm and 6.5 nm$^2$ after the structural relaxation; the number of basic building blocks is 1,224. Both the cases of SiOC(H) molecule before and after UV treatment are simulated by the commercial MD solver Discover (version 2005.2) [13], and the force fields between the atoms are described by COMPASS (definition: cff91, version 2.6). Both computations are performed on an i686 machine with 2.8 GHz CPU and CPU time for each case is approximately 270,500 s. In this paper, the canonical ensemble (NVT) ensemble, which conserves the number of atoms ($N$), the system volume ($V$) and the temperature ($T$), is used. Moreover, no periodic boundary condition is applied to any model.

The simulation results are listed for case A1 and A2 in Table 8.1. The simulation shows that the Young's modulus and density of A2 (after UV treatment) is slightly higher than A1 (before UV treatment), and the similar trend is also found in the experiment [4]. Note that the simulated density is defined as the ratio of atomic mass and molecule volume. Note that the molecule volume is defined as the volume which is occupied by the molecular surface. This simple case study

**Fig. 8.9** The molecular model A1. **a** Side view, **b** cross section view

**Table 8.1** Parametric analysis of the SiOC(H)

| Case | Ratio of basic building blocks | | | | | Young's modulus |
|------|---------|---------|---------|---------|-----------|---------|
| | $r_Q$ (%) | $r_T$ (%) | $r_D$ (%) | $r_M$ (%) | $r_{Void}$ (%) | (GPa) |
| A1 | 16 | 44 | 29 | 1 | 10 | 16.38 |
| A2 | 18 | 43 | 16 | 3 | 20 | 22.53 |
| B1 | 21 | 39 | 29 | 1 | 10 | 17.87 |
| B2 | 31 | 29 | 29 | 1 | 10 | 22.80 |
| B3 | 15 | 45 | 16 | 3 | 21 | 16.07 |
| C1 | 70 | 24 | 6 | 0 | 0 | 43.85 |
| C2 | 22 | 68 | 9 | 0 | 1 | 24.93 |
| C3 | 11 | 19 | 69 | 0 | 2 | 11.46 |
| D1 | 0 | 0 | 0 | 0 | 100 | 0.0 |

demonstrates that the MD simulation has the capability to describe the variation of Young's modulus and density as function of chemical composition.

## 8.4.1 Data Management of MD Simulation Results and Experimental Validation

In order to verify the accuracy of the amorphous structure generated by the proposed method, two molecules, the SiOC:H before and after UV treatment, are used as the qualitative validation. In Ref. [4], the concentrations of the basic building blocks and the Si–O–Si bond angles of two molecules have been measured by nuclear magnetic resonance (NMR) and Fourier transform infrared spectroscopy (FTIR), respectively. FTIR results indicated that the SiOC:H will lose the large angle Si–O–Si bond angles after the UV treatment. Two models, based on the experimentally obtained building block concentrations, are established by the proposed method. The bond angle distributions of these two models confirm the loss of large angles, which is shown in Fig. 8.10.

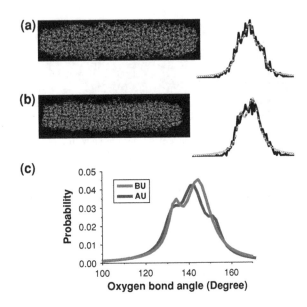

**Fig. 8.10** Qualitative validation of the obtained SiOC:H structure by oxygen bond angle distribution (BAD). *Left hand sides* of **a** and **b** represent the model before and after UV treatment. *Right hand sides* of **a** and **b** are the oxygen BAD of BU and AU (*thick black line*) and their Lorenzian fitting (*red dash line with hollow dots*). **c** is the comparison of oxygen BAD fitting, and the maximum Oxygen bond angle shifts to *left hand side* after UV treatment

**Table 8.2** Experimental validation on predicted trend

| SiOC(H) | | BU[a] | AU[a] |
|---|---|---|---|
| Concentration[b] | Q | 15.70% | 21.70% |
| | T | 47.40% | 49.70% |
| | D | 29.80% | 20.50% |
| By Fitting function[c] | E (GPa) | 18.3 | 21.9 |
| | D (g/cm³) | 1.81 | 1.90 |
| By Experiment[d] | E (GPa) | 11 ± 1 | 16 ± 1 |
| | D (g/cm³) | 1.48 | 1.52 |

[a] : BU and AU represent the SiOC(H) molecule before and after UV treatment
[b] : obtained by nuclear magnetic resonance (NMR)
[c] : obtained by nano indenter
[d] : obtained by X-ray reflectivity (XRR)

The simulation shows that the Young's modulus and density of A2 (after UV treatment) is slightly higher than A1 (before UV treatment)—comparable to a similar trend also found in the experimental results listed in Table 8.2. Note that the simulated density is defined as the ratio of atomic mass and molecule volume, where the molecule volume is defined as the volume which is occupied by the molecular surface with the Connolly radius of 0.1 nm [13]. This simple case study demonstrates that the MD simulation is capable of describing the variation of Young's modulus and density as function of chemical composition.

In order to understand how the concentration of Q, T, D and void impact the Young's modulus and density, A response function,

**Table 8.3** Fitting coefficients

|                        | $c_0$ | $E_Q$ | $E_T$ | $E_D$ |
|------------------------|-------|-------|-------|-------|
| Young's modulus (GPa)  | 0.25  | 55.78 | 18.42 | 1.76  |

$$f_{E,density} = c_0 + c_Q r_Q + c_T r_T + c_D r_D + c_{void} r_{void} \qquad (8.1)$$

is used to obtain the sensitivity of the parametric analysis, listed in the Table 8.2. For simplification, the ratio of M is merged into D because the ratio of M is relatively small compared to the rest.

Moreover, a rather simple fitting function based on homogenization theory is used to describe the numerical results. We denote Young's moduli and densities of 100% Q, T, D are $E_Q$, $E_T$, $E_D$, $P_Q$, $P_T$ and $P_D$, respectively. Hence two fitting functions for Young's modulus and density can be written as:

$$E = E_Q r_Q + E_T r_T + E_D r_D \qquad (8.2a)$$

$$\rho = \rho_Q r_Q + \rho_T r_T + \rho_D r_D \qquad (8.2b)$$

The coefficients can be obtained by the least square method (Table 8.3). Considering the detail experimental data on SiOC(H) molecule before and after UV treatment, the concentrations of basic building blocks are shown in Fig. 8.11. The root mean square differences between the fitting function and each cases for Young's modulus and Density are 0.54 (GPa) and 0.033 (g/cm$^3$), respectively, shown in Fig. 8.12. The fitting result shows that the Young's modulus of the SiOC:H is dominated by the concentration of basic building block Q and T, and the concentration of the basic block.

Notably, the obtained, which also can be theoretically interpreted as the Young's modulus of the amorphous SiO$_2$, is 55.78 GPa. Moreover, the experimental results of the amorphous SiO$_2$ films made from tetraethylorthosilane (TEOS) before annealing are approximately 56 and 59 GPa [14, 15], which have been measured by the Brillouin light scattering (BLS) technique and bending beam method, respectively.

## 8.4.2 Interface Modeling by Molecular Dynamics

Considering a system with an interface and a set of external loading applied onto that system. Assume that interfacial failure is observed by experiments. Therefore, according to the conservation of the total energy of the system, the work which is done by the external loading will equal the summation of material deformation energy, heat and the surface energy (intrinsic adhesion) of the interface [16]. Theoretically speaking, the intrinsic adhesion of the interface consists of three

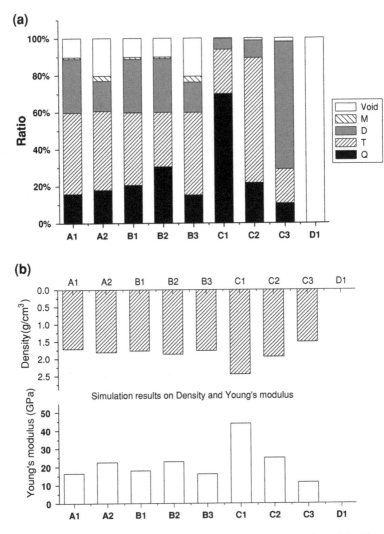

**Fig. 8.11** The plots of concentration of basic blocks, density and Young's modulus (from *upper panel* to *lower panel*)

potential contributors: chemical interaction, physical interaction and mechanical interlocking.

- Chemical interactions: The chemical interactions between the interfaces often refer to the covalent bond, ionic bond or the metallic bond. These bonds are relatively strong and termed as primary bonds. In the electronic or packaging laminated structure, the most common chemical interaction at the interface is

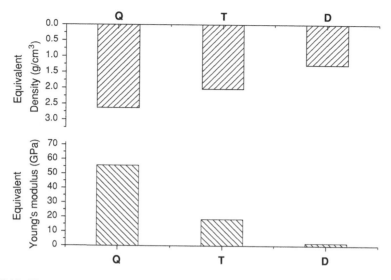

**Fig. 8.12** The sensitivity of Young's modulus and density

the covalent bonds besides the alloy system. The interaction scale of the chemical interaction is approximately 0.2–1.0 nm.

- Physical interactions: The physical interaction often refers to the weak bonds between interface, like the Coulomb force or van der Waals force. Although the magnitude of the physical interaction is weaker than the one of the chemical bond, the physical can be formed at most interfaces and chemical interaction requires certain chemical condition to form. The interaction scale of the physical interaction is approximately 5.0–10.0 nm. Moreover, considering a polymer interface system, two chained polymers can entangle together when the polymer chains have enough energy to move though the interface and this phenomenon also contribute to the interfacial strength of polymer and polymer system.
- Mechanical interlocking: Both the physical and chemical adhesion mechanisms are on the microscopic scale. At the macroscopic scale, mechanical interlocking can be applied, as in the surface treatments applied to metal to increase the surface roughness and to obtain better adhesion strength. For a specific process and materials, the pattern and distribution of the surface roughness can be controlled. The interaction scale of the physical interaction is approximately larger than 10.0 nm.

Within the low-$k$ material interfacial modeling, before the interfacial strength simulation, the situation of the chemical bond at the interface should be well defined. There is few theory or experimental method which is capable of predicting/measuring the chemical configuration at the interface. Therefore, to simulation the mechanical de-lamination at molecular level, an engineering approach is proposed. This approach use the artificial charge to attract two molecules and the

**Fig. 8.13** Illustration of the
covalent bond generating
algorithm for SiOC:H/Silica
interface. In **a–d**, the *upper
panel* is amorphous Silica and
the lower panel is SiOC:H.
The chemical bonds of the
interfacial atoms are
illustrated but not the interior
atoms, which are surrounded
by the *dashed lines*. **a** The
initial status. **b** Removed and
charged interfacial atoms.
**c** Molecules attract each
other. **d** Calculate the
distance of interfacial atoms

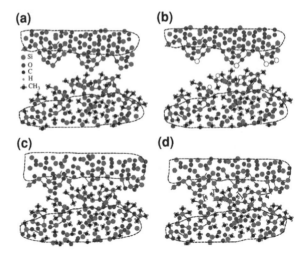

covalent bond is established when the distance between two atoms is a satisfied predefined criterion. For this low-$k$ to silica interface (Fig. 8.13a) which the chemical bond can be formed (judged by chemical reaction energy), an engineering approach is applied as following:

1. The silicon atoms at silica interface and oxygen atoms at low-$k$ interface are artificially removed.
2. The remaining oxygen and silicon atoms at interface are artificially charged. A positive charge is applied to the silicon atom and negative to the oxygen (Fig. 8.13b).
3. A molecular dynamics simulation is performed and the two separated molecules are attracted to each other by the electrostatic force (Fig. 8.13c).
4. When the distance of the atoms at interface smaller than the criterion, the chemical bond is formed. Afterward, the artificial charges are removed, and the partial charges of the atoms are re-calculated (Fig. 8.13d). This algorithm can provide a scheme of covalent bond distribution at interface. The forming criterion at step 4 can be a tuning parameter with respect to the covalent bond density along the interface.

Figure 8.14a shows the molecular model of TEOS/low-$k$. A pre-described displacement is applied to the top few layer of atoms with fixed velocity. The bottom few layer of atoms is fixed and the reaction force can be obtained.

Based on the applied displacement and obtained reaction force, the energy of interfacial fracture can be obtained in Fig. 8.14b. The area under Fig. 8.14b is considered to be the external energy to fracture (fracture energy) the structure, with the calculation of 19.7 nN-nm. Consider the fracture energy per area, we have 1.87 N/m (with the surface of 10.53 nm$^2$, and Connolly radius of 1.0 nm). The experimental results of BD2/TEOS interface from the four-point-bending method

**Fig. 8.14** MD modeling technology: **a** is the MD model of the TEOS/low-$k$ interface, where the low-$k$ material is Black Diamond (BD). **b** shows simulation result of applied displacement and reaction force

**(a)**

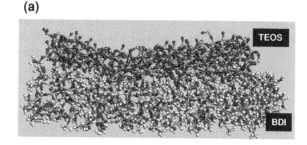

**(b)**

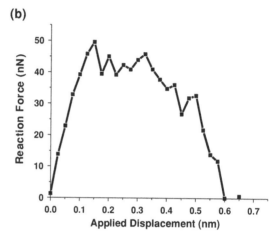

gives the fracture toughness of approximately 1.9–2.7 (N/m) [17], as pure mode I (opening) fracture. The simulation result is slightly lower the experimental one, due to the mechanical interlocking is not well-described in the model.

## 8.5 Conclusions

In this paper, a prediction method for mechanical/interfacial strengths of amorphous material is presented. The molecular dynamics (MD) method is used because the atomic structure can be described as well as the interaction force between atoms. Before the simulation of the interfacial strength, an engineering approach is applied to model the chemical configuration at the interface. The same simulation procedure, which used in the fracture strength simulation, is applied to obtain the interfacial strength as well as the fracture (delaminated) energy. Moreover, the unique modeling technique for the atomic status at molecular interface is presented, and the interfacial strength can be obtained. The simulation results also imply that the existence of the covalent bond has a significant influence to the interfacial system in the silica/low-$k$ system. Moreover, the delamination is not always due to the covalent bonds at interface, but the failure in the soft material (e.g., SiOC:H). The simulation results indicate that the crack will be initialized around the pore in the low-$k$ material. Therefore, the interfacial strength can be improved by enhancing local stiffness of few nm thickness of the low-$k$ material near the interface and preventing the pore near the interface. Increasing the amount of covalent bond at interface will contribute less, because the crack will propagate along the weakest part of the interfacial system. However, there is an enormous challenge to predict the initiation and propagation of an interfacial system in the nano-scaled, due to unknown of defects of the material and interface, lack of sufficient force field, lack of robust multi-scaled modeling technique and lack of exact chemical configuration of interface.

# References

1. International Technology Roadmap for Semiconductors, ITRS, 2006 (updated)
2. Grill A, Neumayer DA (2003) Structure of low dielectric constant to extreme low dielectric constant SiCOH films Fourier transform infrared spectroscopy characterization. J Appl Phys 94(10):6697–6707
3. Maex K, Baklanov MR, Shamiryan D, Iacopi F, Brongersma SH, Yanovitskaya ZS (2003) Low dielectric constant materials for microelectronics. J Appl Phys 93(11):8793–8884
4. Iacopi F, Travaly Y, Eyckens B, Waldfried C, Abell T, Guyer EP, Gage DM, Dauskardt RH, Sajavaara T, Houthoofd K, Grobet P, Jacobs P, Maex K (2006) Short-ranged structural rearrangements and enhancement of mechanical properties of organosilicate glasses induced by ultraviolet radiation. J Appl Phys 99:053511
5. Falk ML, Langer JS (1998) Dynamics of viscoplastic deformation in amorphous solids. Phys Rev E 57:7192–7205
6. Yuan CA, van der Sluis O, Zhang GQ, Ernst LJ, Flower A E, van Silfhout RBR, Thijsse BJ (2007) Chemical-mechanical relationship of amorphous, porous low-dielectric film materials. Comput Mater Sci
7. Yuan CA, van der Sluis O, Zhang GQ, van Driel WD, Ernst LJ, van Silfhout RBR (2007) Molecular simulation on the material/interfacial strength of the low-dielectric materials. Microelectron Reliab 47:1483–1491
8. Yuan CA, van der Sluis O, Zhang GQ, Ernst LJ, van Driel WD, Flower AE, van Silfhout RBR (2008) Molecular simulation strategy for the amorphous/porous low-dielectric constant materials. Appl Phys Lett 92:061909
9. Yuan CA, van der Sluis O, van Driel WD, Zhang GQ (2008) The need for multi-scale approaches in Cu/low-$k$ reliability issues. Microelectron Reliab 48:833–842
10. Love AEH (1934) A treatise on the mathematical theory of elasticity. Cambridge University Press, New York
11. Hod O, Scuseria GE (2008) Half-metallic zigzag carbon nanotube dots. ACS Nano 2:2243–2249
12. Zhang YY et al (2009) Buckling of carbon nanotubes at high temperatures. Nanotechnology 20:215702
13. Accelrys Inc (2005) Materials Studio$^{TM}$ DISCOVER. Accelrys Inc., San Diego
14. Carlotti G, Colpani P, Piccolo D, Santucci S, Senez V, Socino G, Verdini L (2002) Measurement of the elastic and viscoelastic properties of dielectric films used in microelectronics. Thin Solid Films 414:99–104

15. Zhao J, Ryan T, Ho PS, McKerrow AJ, Shih W (1999) Measurement of elastic modulus, Poisson ratio, and coefficient of thermal expansion of on-wafer submicron films. J appl phys 85:6421–6424
16. Ernst LJ, van Driel WD, van der Sluis O, Corigliano O, Tay AAO, Iwamoto N, Yuen MMF (2006) Fracture and delamination in microelectronics. In: Proceedings of the Asian–Pacific conference for fracture and Strength (APCFS'06), Singapore
17. Kouters MHM (2006) Characterisation of interfacial strength of low-$k$ dielectric materials used in ICs. MSc thesis (MT 06.37), TU Eindhoven

# Chapter 9
# Molecular Design of Self-Assembled Monolayer (SAM) Coupling Agent for Reliable Interfaces by Molecular Dynamics Simulation

**C. K. Y. Wong, H. Fan, G. Q. (Kouchi) Zhang and M. M. F. Yuen**

This paper is based upon "Molecular design of reliable Cu–epoxy interface using Molecular Dynamic Simulation", by C.K.Y. Wong, H.B. Fan, G.Q. Zhang and M.M.F. Yuen which appeared in the Proceedings of Eurosime 2010 © Year, IEEE.

**Abstract** This paper aims at developing molecular modeling methodology to select a thiol-based self assembly monolayer (SAM) as a coupling agent for achieving a reliable epoxy–copper interfacial adhesion under moisture conditions. Moisture diffusion and interfacial energy is evaluated using molecular dynamics simulations. The qualitative agreement of the calculated interfacial energy with the experimental adhesion energy demonstrates that the molecular dynamics method is an effective way in selecting the coupling agent candidates.

## 9.1 Introduction

Epoxy–copper (epoxy–Cu) interface is known to be one of the weakest interfaces in an electronic package which exhibits failure such as delamination during reliability tests. Previous studies showed serious degradation of the interfaces

C. K. Y. Wong (✉) · G. Q. (Kouchi) Zhang
Faculty 3mE, Department PME, Delft University of Technology,
Mekelweg 2, 2628 CD Delft, The Netherlands
e-mail: K.Y.Leung-Wong@tudelft.nl

H. Fan · M. M. F. Yuen
Department of Mechanical Engineering, Hong Kong University of Science and Technology, Clear Water Bay, Kowloon, Hong Kong (SAR), China

N. Iwamoto et al. (eds.), *Molecular Modeling and Multiscaling Issues for Electronic Material Applications*, DOI: 10.1007/978-1-4614-1728-6_9,
© Springer Science+Business Media, LLC 2012

under moisture sensitive conditions. Kim et al. [1] reported the button shear strength of black oxide treated interface. The interfacial adhesion decrease by 83.1% from 7.5 to 1.27 MPa was reported in a pressure cooker test (121°C/100%RH for 120 h). With the oxide samples pre-conditioned under 85°C/85%RH, 168 h, Takano et al. [2] demonstrated a decrease of 41.7% in a button shear test. The study indicates inefficient moisture resistance of the black oxide treatment to the interface.

Several groups attribute the cause of interfacial delamination in an epoxy–Cu joint to moisture uptake of the epoxy polymer and relate the adhesion to moisture diffusion along the interface. Soles et al. [3] recommended the hydrophobic component in an epoxy system that can hinder moisture uptake of the bulk epoxy. Kinloch [4] suggested that the long alkyl chain of a vinyl silane coupling agent with 20 carbons might impede water and improve bond durability. The argument is the middle alkyl chains, which are highly hydrophobic, can, obstruct water penetration and can thus improve the interface. These studies suggest that introducing hydrophobic characteristic in the epoxy–Cu interface may reduce the moisture content at the interface. The introduction of hydrophobicity at an interface can inhibit interfacial moisture diffusion and improve the long-term reliability of an interfacial joint.

One way to establish hydrophobicity is to modify a substrate with coupling agent having hydrophobic characteristics. Despite the ease in having coupling agent with hydrophobicity, selection of the appropriate candidate is a headache with tremendous experimental effort. Fan et al. suggest to use molecular dynamics (MD) simulation for molecular design of coupling agent by studying the interfacial energy of the system. The adhesion of an interface is evaluated by subtracting the potential energy of the system with that of the components (copper and epoxy in this case) [5]. Qualitative agreement has been obtained from MD calculation and the experiment. As discussed, interfacial moisture diffusion may have adverse effect on interface reliability; determination of interfacial moisture diffusion is worthy but difficult. Several references have reported the calculation of moisture diffusion in polymer by MD simulations [6–9]. Calculation of moisture diffusion coefficient at an interface is also possible by MD models [10, 11].

This study aims at using molecular dynamics simulation to select a coupling agent candidate to solve the reliability problem of epoxy–copper interface. Thiol-based self assembly monolayer (SAM) has been chosen as the coupling agent for its ready reaction with copper and the ease of properties modification with the organic function group. The moisture diffusion behavior of a modified interface is evaluated by calculating the diffusion coefficient of water molecules along the SAM modified interfaces. The interfacial adhesion of the SAM modified system in moisture sensitive condition is assessed by interfacial energy calculation. The value is then benchmarked experimentally by measuring the critical energy release rate of the interface under various treatments. This study develops a methodology in using MD simulation in the molecular design of SAM coupling agent for a reliable epoxy–Cu interface.

## 9.2 Molecular Dynamics Simulation

Molecular dynamic (MD) simulation was employed to study the moisture diffusion behavior of the SAM modified epoxy–Cu interfaces and their adhesion property. Figure 9.1 shows the strategy of the modeling.

The simulations were conducted using Material Studio® software developed by Accelrys Software Inc. The condensed phase optimization molecular potentials for atomistic simulation studies (COMPASS) which is commonly adopted to predict properties of polymers, metals and their interfaces [12, 13] over a wide range of temperature and pressure conditions is incorporated to calculate the interaction energy among the systems in our model MD model setup.

In simulating the epoxy–Cu adhesion, molecular models were built according to the designed chemical structures of the SAM candidates and the epoxy adhesive used in this study. Three unit cells namely: substrate, water and epoxy unit cells were built. All the unit cells were in 2.287 nm × 2.287 nm dimensions with periodicity in $x$ and $y$ directions.

The substrate unit cells were constructed with a crystalline copper structure cleaved along (001) plane. On top of the Cu layer, sulphur head of a thiol molecule was assigned to bond with a copper atom. The sulphur atoms were fixed at a position 2.37 Å above the surface copper atoms along the $z$-direction, which was taken from the Cu–S bond length suggested by Jackson et al. [14]. Three types of SAM with different molecular structures were used in this study. Models were built according to the structures. It assumes that only one layer of thiol molecules was bonded to the copper substrates. To assign the number of SAM molecules on the substrate unit cell, a series of MD models comprising SAM chains of different densities were studied. The density ranged from $3.18 \times 10^{-11}$ to $5.72 \times 10^{-10}$ mol/cm$^2$, which corresponded to 1 to 18 SAM chains in a 2.287 nm × 2.287 nm unit cell. MD simulations were performed to find out the most stable structure for a given configuration. Conformation was carried out at 25°C in a canonical (NVT) ensemble. The atom-based summation with a 9.5 Å cutoff distance was used in evaluating the non-bonded interactions. The system was brought to equilibrium for 20 ps with 1.0 fs time step. The calculated potential energy of a structure represents its thermodynamic state. A structure having the lowest potential energy implies that it is under the most stable state and it is the most statistically possible structure one could obtain at that condition. For each of the evaluated SAM structures, the SAM density with the lowest potential energy would be used in the later section for model construction.

Epoxy unit cell was built according to the major components of the epoxy adhesive used in the benchmarking experiment. An epoxy chains, which originated from cycloaliphatic epoxy, anhydride and bisphenol A, was constructed in an amorphous cell with a density of 1.7 g/cm$^3$ in accordance with the typical density of the commercial adhesive. A picture showing the schematic for the chemical structure of the epoxy unit cell is given in Fig. 9.2.

**Fig. 9.1** Modeling strategy

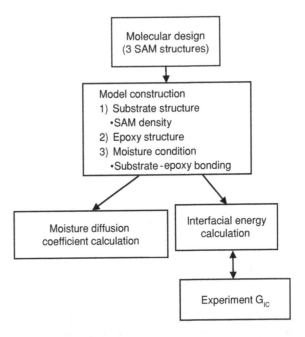

**Fig. 9.2** An epoxy unit cell

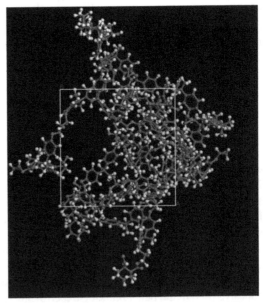

The water unit cell was constructed to investigate the diffusion behavior of water along the interface. A layer of water molecules having 3 wt% of water with respect to the molecular weight of the epoxy was built.

The three unit cells were then combined using a layer builder. A vacuum spacer in the $z$ direction was positioned on top of the epoxy chains. The total thickness of the system was set at 13.7 nm.

With the layer unit cells, the MD models for different SAM modified interfaces are ready for bond assignment. All the candidates are assumed to be bonded with epoxy. Bonding with the epoxy marcomolecules depend on the chemistry of functional groups in the SAM molecules. The number of the epoxy bonding is determined from the energy minimum of the model after optimization. All atoms, except the sulphur and copper atoms, were allowed to move freely. Canonical (NVT) ensemble was adopted in the simulation at a temperature of 80°C according to the epoxy curing condition. The systems were equilibrated for 20 ps with 20,000 iteration steps at a time step of 1.0 fs. A 9.5 Å cutoff distance was used in calculating the non-bonded interactions of the models.

## 9.2.1 Moisture Diffusion and Interfacial Failure Model Descriptions

The moisture diffusion simulations were performed at a temperature of 85°C using the isothermal-isobaric ensemble (NPT). Non-bond interactions with cutoff distance of 9.5 Å is adopted. The calculation in each model is performed with an interval of 1 fs with 50,000 time steps. The mean squared displacements of all water molecules were evaluated in each time step to track the motion of the molecules. Diffusion coefficient of the SAM modified interfaces can be obtained in principle from diffusion trajectories $r(t)$ of water molecules determined during an MD simulation of a polymer packing model [15]. The diffusion coefficients for the water molecules could then be calculated from the mean squared displacement, $\langle [r(t) - r(0)]^2 \rangle$, of the water molecules averaged over time as follows:

$$D = \frac{1}{6N} \lim_{x \to \infty} \frac{d}{dt} \sum_{i=1}^{N_x} \left\langle [r_i(t) - r_i(0)]^2 \right\rangle \tag{9.1}$$

where $D$ is the moisture diffusion coefficient, $r_i(t)$ is the coordinate of the center of the mass of the $i$th water molecule and $N$ is the number of water molecules in the system interfacial energy calculation.

Interaction energy of the components in the system can be evaluated by calculating potential energy within the system. The potential energy was calculated by superposition of bonding and non-bond interactions. The bonding terms consist of bond stretch, bond angle bending, dihedral angle torsion energies and the cross term which describes the interaction for bond/angle distortion caused by nearby atoms. The non-bond interactions consist of van der Waals forces and electrostatic energy.

Failure in the models was assumed to occur at an interface with lower strength. The assumption is evolved from the fact that the interface, in a bi-material system, is less able to carry loading and thus more prone to failure. Due to high bond

**Fig. 9.3** The failure
locations assumed in the MD
models for the epoxy–SAM–
Cu system

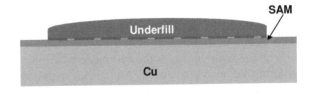

energy of the Cu–S bond (276 kJmol$^{-1}$), failure is not expected at the Cu–SAM interface. It is therefore assumed that failure occurs at the epoxy–SAM interface which it is not instigated within the bulk materials. Figure 9.3 depicts the assumed failure location in the models.

To obtain the interfacial energy of SAM modified interfaces under moisture condition, the system was equilibrated at canonical (NVT) ensemble for 20 ps at 25°C. The simulation was operated with 20,000 iteration steps with a time step of 1.0 fs and 9.5 Å cutoff distance in calculating the non-bonded interactions.

The interfacial energy density, also known as interaction energy, was calculated as Eq. 9.2:

$$\Delta E = E_{\text{total}} - (E_1 + E_2) \tag{9.2}$$

where $E_{\text{total}}$ is the total potential energy of the whole system, $E_1$ is the potential energy of the substrate unit cell (includes SAM molecules, copper atoms and water molecules) and $E_2$ is the potential energy for the epoxy layer. The interfacial energy (in Jm$^{-2}$) for a given interface per unit area, $\gamma$, was evaluated by Eq. 9.3.

$$\gamma = \frac{\Delta E}{A} \tag{9.3}$$

where $A$ is the contact area for interaction between the substrate and epoxy in the model.

## 9.2.2 Benchmarking Experiment

The interfacial adhesion of the epoxy–Cu joint was evaluated in terms of the critical energy release rate ($G_{\text{IC}}$) of the modified joint. Tests were performed with Tapered Double Cantilever Beam (TDCB) specimens in accordance with ASTM D3433. The specimen was prepared from epoxy adhesive sandwiched between a pair of tapered Cu substrates as illustrated schematically in Fig. 9.4. The tapered geometry guarantees a constant moment during the crack growth. In order to eliminate the roughness effect, the tapered Cu substrates was polished with P4000 sandpaper before surface treatment. To ensure that crack opened at the right interface, a 60 mm precrack was introduced at the front end of the substrate.

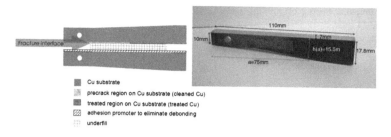

Cu substrate
precrack region on Cu substrate (cleaned Cu)
treated region on Cu substrate (treated Cu)
adhesion promoter to eliminate debonding
underfill

**Fig. 9.4** Schematic diagram for TDCB configuration

Epoxy adhesive of 0.5 mm thickness was cured between a pair of copper substrates at 80°C for 6 h. The width and total length of the substrate was 7 and 102 mm, respectively.

The bonded specimen was tested in an MTS 858 Universal Testing Machine equipped with a 25 kN ± 5 N axial force load cell. Crosshead displacement rate was kept constant as 30 µm/min in tension under control displacement mode. It was assumed that the crack developed under linear fracture mechanics condition. The $G_{IC}$ was calculated from Eq. 9.4.

$$G_{IC} = \frac{4P^2 [3a^2 + h(a)^2]}{EB^2 h(a)^3} \tag{9.4}$$

where $P$ is the critical load for crack opening; $B$ is the width of the specimen; $a$ is the crack length and $h$ is the corresponding thickness of the substrate at the point of crack front and $E$ is Young's modulus of the substrate. For a given geometry, $m$ is a constant for the tapered substrate,

$$m = \frac{3a^2 + h(a)^2}{h(a)^3} \tag{9.5}$$

The reported $G_{IC}$ was averaged from no less than four specimens for every set of SAM candidates. Both freshly prepared samples and the samples after aging at a 85°C/85%RH humidity chamber were performed to assess the moisture resistance properties of the interface.

## 9.3  Results and Discussion

### 9.3.1  SAM Density Calculation

Figure 9.5 shows the potential energy of a substrate unit cell against SAM density for the SAM candidates. The potential energy is given as the energy of the substrate unit cell for a model structure. The potential energy was calculated to

**Fig. 9.5** Potential energy of
various SAM treated
substrate against SAM
density

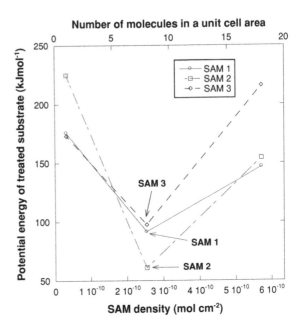

cover the cases for one, eight and 18 SAM molecules attached to the copper atoms. These numbers are the allowable configuration in maintaining symmetry about the x and y axes. All the candidates show the energy minimum when eight molecules are put inside the unit cell. The density for the most stable structures (with the lowest potential energy) was estimated as $2.54 \times 10^{-10}$ mol/cm$^2$ which corresponded to 65.4 Å$^2$/molecule for all the SAM structures.

### 9.3.2 Number of Epoxy Bonds

To simulate the covalent bonding effect in the models, epoxy rings in an epoxy unit cell were connected to functional groups in the SAM molecules. Prior to covalent bonding, six unreacted rings were contained in the epoxy unit cell. However, not all of these rings should be bonded to the SAM molecules. If distant epoxy rings were connected to a SAM molecule, large strain energy would be induced making the system unstable. To avoid instability, only one to five epoxy rings were bonded to the SAM. The potential energy for those configurations was calculated. The configuration with the minimum total potential energy was chosen as the bond assignment scheme for this study.

Figure 9.6 shows the change in the potential energy of model systems as the substrate was bonded with different numbers of epoxy rings. Energy minimum was achieved with two epoxy rings bonded to both the SAM A and SAM B substrates

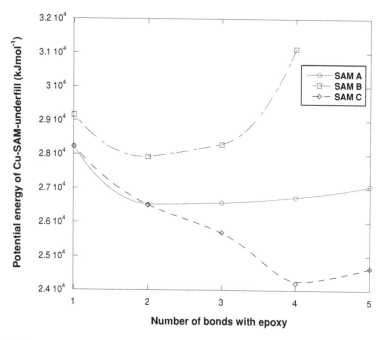

**Fig. 9.6** The total potential energy of the Cu-SAM-EMC systems for different SAM systems against number of bonds with epoxy

while four epoxy bonds gives the most stable configuration for SAM C. The energy increased significantly as the epoxy chains were stretched with increasing strain to bond with the substrates. The building of the MD models is thus done according to these bond assignments.

After the combination of the unit cells and the bond assignment, conformation of the models was performed. Figures 9.7, 9.8, and 9.9 illustrates the final configuration of the model for different interfaces under various SAM candidate treatment.

### 9.3.3 Moisture Diffusion Calculation

Figure 9.10 demonstrates the mean square displacement of the water molecules against time at different SAM interfaces obtained from the simulations. The graph was fitted using linear regression, $y = b + ax$ with the slope, a. Using Eq. 9.1, the moisture diffusion coefficient at the interfaces in different SAM treatments is calculated and reported in Table 9.1.

These values represent water penetrates most actively at epoxy–SAM A–Cu interface while slower water molecules transportation in SAM C treated interface.

**Fig. 9.7** Morphological configuration of SAM A treated epoxy–Cu interfaces under moisture conditions

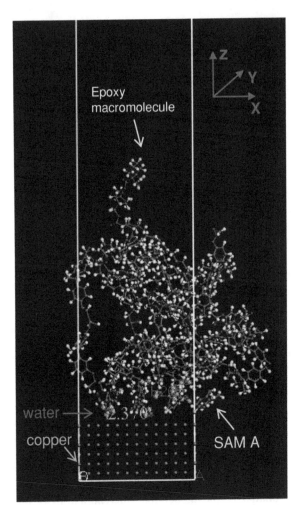

## 9.3.4 Interfacial Energy Calculation

The interfacial energy for the epoxy–Cu systems can be calculated using the potential energy of a system in the simulations. The interaction energy between the substrate and epoxy was calculated according to Eqs. 9.2 and 9.3.

Table 9.2 presents the calculated interfacial energy for the systems. The interfacial energy is 1.56, 2.72 and 1.24 $Jm^{-2}$, respectively for the SAM A, SAM B and SAM C systems. The data imply that, SAM B coupling agent is expected to result in highest adhesion among the candidates.

**Fig. 9.8** Morphological configuration of SAM B treated epoxy–Cu interfaces under moisture conditions

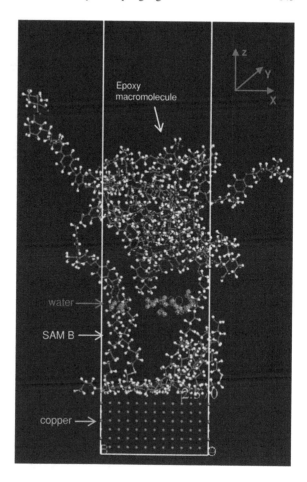

### 9.3.5 TDCB Benchmarking

The interfacial adhesion is evaluated with the TDCB specimens. The critical energy release rate ($G_{IC}$) of the surface modified samples was calculated from Eq. 9.4. Figure 9.11 gives the $G_{IC}$ of the freshly prepared and 85°C/85%RH aged samples with different SAM treatment. Without any treatment, the $G_{IC}$ of the freshly prepared samples was 4.8 $Jm^{-2}$ [16]. Upon SAM treatment, the $G_{IC}$ increase to $159 \pm 33$, $118 \pm 29$ and $46 \pm 6$ $Jm^{-2}$ for SAM A, SAM B and SAM C, respectively. Despite the best result obtained from SAM A in the fresh sample, sufficient decrease in adhesion has been recorded for this treatment in aging to $93 \pm 16$ $Jm^{-2}$. The $G_{IC}$ was almost unchanged with the SAM B treatment while only a little drop was measured with SAM C. The results imply that SAM A makes the most moisture sensitive interface with poor reliability.

**Fig. 9.9** Morphological
configuration of SAM C
treated epoxy–Cu interfaces
under moisture conditions

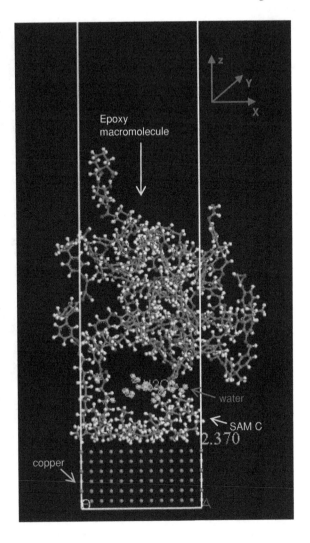

A plot illustrates the $G_{IC}$ obtained from TDCB experiment against interfacial energy calculated from simulation is shown in Fig. 9.12. Although the value estimated from MD is significantly smaller than the experimental data, a similar trend is obtained.

Table 9.3 summarizes the moisture diffusion coefficient and the percentage decrease in adhesion of the SAM treated interface. The adhesion deterioration upon moisture pre-conditioning is reported as the percentage decrease in adhesion as calculated from Eq. 9.6. The percentage decrease for SAM A, SAM B and SAM C are 41.7, 0 and 12.4%, respectively. The adhesion deterioration for SAM A is comparable with those with copper oxide treatment [2]. SAM B and SAM C show much less deterioration as the prediction from the simulation.

**Fig. 9.10** Mean square displacement of water molecules against time and the fitted line

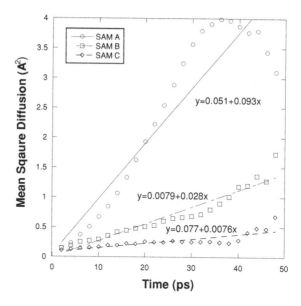

**Table 9.1** The moisture diffusion coefficient, D of epoxy–Cu interface in different SAM treatment

|       | Slope, a ($\times 10^{-2}$ mm$^2$ s$^{-1}$) | Moisture diffusion coefficient (mm$^2$ s$^{-1}$) |
|-------|------------------------|-----------------------------------|
| SAM A | 0.093                  | $1.29 \times 10^{-5}$             |
| SAM B | 0.028                  | $3.90 \times 10^{-6}$             |
| SAM C | 0.0076                 | $1.06 \times 10^{-6}$             |

**Table 9.2** MD results for interfacial energy density of the three models

|       | Interfacial energy per unit cell (kJ mol$^{-1}$) | Interfacial energy per meter square (Jm$^{-2}$) |
|-------|------------------------|-----------------------------------|
| SAM A | −4905                  | 1.56                              |
| SAM B | −8555                  | 2.72                              |
| SAM C | −3915                  | 1.24                              |

$$\%_{\text{decrease}} = \frac{G_{\text{IC,inital}} - G_{\text{IC,aged}}}{G_{\text{IC,inital}}} \tag{9.6}$$

Nevertheless, the data show that the decrease in moisture diffusion may not directly result in improvement in adhesion degradation. The phenomenon may be explained by:

1. The assumption of 3 wt% moisture content at the interface may be different for different SAM treatment. Rather than obtaining from the bulk moisture content, experiments should be done to verify this value.
2. The degradation may not only result from moisture diffusion, it may also be due to weakening of bonds which affect the interaction at the interface.

**Fig. 9.11** Critical energy
release rate ($G_{IC}$) of epoxy–
Cu interface prepared from
different SAM treatment with
both fresh and 85°C/85%RH
aged conditions

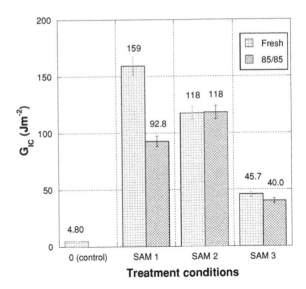

**Fig. 9.12** The interfacial
energy predicted by MD
against the measurement
value from the TDCB test

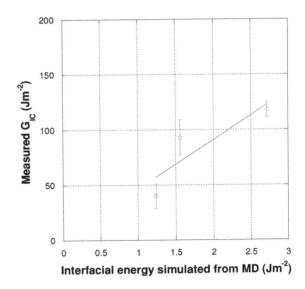

**Table 9.3** Percentage
decrease in adhesion upon
aging

|  | Moisture diffusion coefficient (mm² s⁻¹) | Percentage decrease (%) |
|---|---|---|
| SAM A | $1.29 \times 10^{-5}$ | 41.7 |
| SAM B | $3.90 \times 10^{-6}$ | 0 |
| SAM C | $1.06 \times 10^{-6}$ | 12.4 |

## 9.4 Summary

This work demonstrates a methodology in using MD simulation to select SAM candidates for a reliable epoxy–Cu interface. The qualitative agreement of the interfacial energy calculated from simulations and the measured $G_{IC}$ makes MD an efficient method in molecular design of coupling agents. The moisture diffusion analysis shows the importance in obtaining the experimental diffusion data to input to the MD model. Further work needs to be conducted to understand the interfacial diffusion phenomenon.

**Acknowledgments** The work was supported by HKSAR under General Research Fund 621907 and Innovation and Technology Fund ITS/331/09. The authors also acknowledge the support from the European Commission under the Nano Interface Project NMP-2008-214371.

## References

1. Kim JK, Lebbai M, Lam YM, Hung PYP, Woo RSC (2005) Effects of moisture and temperature ageing on reliability of interfacial adhesion with black copper oxide substrate. J Adhes Sci Technol 19:427–444
2. Takano E, Mino T, Takahashi K, Sawada K, Shimizu SY, Yoo HY (1997) Oxidation control of copper leadframe package for prevention of popcorn cracking. In: 47th Electronic components and technology conference, pp 78–83
3. Soles CL, Chang FT, Bolan BA, Hristov HA, Gidley DW, Yee AF (1998) Contributions of the nanovoid structure to the moisture absorption properties of epoxy resins. J Polym Sci Part B Polym Phys 36:3035–3048
4. Kinloch AJ, Tan KT, Watts JF (2006) Novel self-assembling silane for abhesive and adhesive applications. J Adhes 82:1117–1132
5. Wong CKY, Fan H, Yuen MMF (2005) Investigation of adhesion properties of Cu–EMC interface by molecular dynamic simulation. In: Proceedings of the 6th international conference on thermal, mechanical and multi-physics simulation and experiments in microelectronics and micro-systems—EuroSimE 2005, Berlin, pp 31–35
6. Müller-Plathe F, Rogers SC, van Gunsteren WF (1992) Computational evidence for anomalous diffusion of small molecules in amorphous polymers. Chem Phys Lett 199:237–243
7. Hofmann D, Fritz L, Ulbrich J, Schepers C, Böhning M (2000) Detailed atomistic molecular modeling of small molecule diffusion and solution processes in polymeric membrane materials. Macromol Theory Simul 9:293–327
8. Fukuda M, Kuwajima S (1997) Molecular-dynamics simulation of moisture diffusion in polyethylene beyond 10 ns duration. J Chem Phys 107:2149
9. Dermitzaki E, Wunderle B, Bauer J, Walter H, Michel B (2008) Structure property correlation of epoxy resins under the influence of moisture and temperature; and comparison of diffusion coefficient with MD—simulations. In: Proceedings of the 2nd Electronics systemintegration technology conference (ESTC) 2008. Greenwich, pp 897–902
10. Bojan MJ, Steele W (1993) Computer simulations of diffusion in monolayers physisorbed in model pores. In: Materials research society symposium, pp 127–134
11. Milliken JO, Zollweg JA, Bobalek EG (1980) Interfacial mass transfer-permeation of hard spheres through hard disc monolayers. J Colloid Interface Sci 77:41–49
12. Materials Studio (2003) Accelrys Inc., San Diego

13. Prathaba B, Subramanianb V, Aminabhavi TM (2007) Molecular dynamics simulations to investigate polymer–polymer and polymer–metal oxide interactions. Polym J 48:409–416
14. Jackson GJ, Woodruff DP, Jones RG, Singh NK, Chan ASY, Cowie BCC, Formoso V (2000) Following local adsorption sites through a surface chemical reaction: $CH_3SH$ on $Cu(111)$. Phys Rev Lett 84:119–122
15. Fan HB, Chan EKL, Wong CKY, Yuen MMF (2006) Investigation of moisture diffusion in electronic packages by molecular dynamics simulation. J Adhes Sci Technol 20:1937–1947
16. Wong CKY, Yuen MMF (2008) Thiol-based chemical treatment as adhesion promoter for Cu-epoxy interface. In: Proceedings of the international conference on electronic packaging technology and high density packaging (ICEPT-HDP) 2008. Pudong, Shanghai

# Chapter 10
# Microelectronics Packaging Materials: Correlating Structure and Property Using Molecular Dynamics Simulations

**Ole Hölck and Bernhard Wunderle**

This paper is based upon the following publications: "Molecular Modeling of Microelectronic Packaging Materials—Basic Thermo-Mechanical Property Estimation of a 3D-cross-linked Epoxy/SiO2 Interface" by O. Hölck, E. Dermitzaki, B. Wunderle, J. Bauer and B. Michel, which appeared in the Proceedings of EuroSimE © 2010, IEEE; "Molecular Dynamics Simulation and Mechanical Characterisation for the Establishment of Structure–Property Correlations for Epoxy Resins in Microelectronics Packaging Applications" by B. Wunderle, E. Dermitzaki, O. Hölck, J. Bauer, H. Walter, Q. Shaik, K. Rätzke, F. Faupel and B. Michel, which appeared in the Proceedings of EuroSimE © 2009, IEEE; "Structure Property Correlation of epoxy resins under the influence of moisture and comparison of diffusion coefficient with MD simulations" by E. Dermitzaki, B. Wunderle, J. Bauer, H. Walter and B. Michel, which appeared in the Proceedings of EuroSimE © 2008, IEEE.

**Abstract** This paper presents a combined experimental and simulative property analysis of cross-linked epoxy resins. Several different molecular modeling methods have been used to evaluate properties ranging from thermo-mechanical to diffusive properties, some of which have been found to produce good agreement between simulation and experiment, whereas others do show qualitative agreement and can thus be useful as a quick tendency assessment. The research has also

---

O. Hölck (✉) · B. Wunderle
Fraunhofer Institute Reliability and Microintegration (IZM),
Gustav-Meyer-Allee 25, 13355 Berlin, Germany
e-mail: Ole.Hoelck@izm.fraunhofer.de

O. Hölck · B. Wunderle
Chair Materials and Reliability of Microsystems,
Chemnitz University of Technology, Reichenhainer Str. 70,
09126 Chemnitz, Germany

B. Wunderle
Fraunhofer Institute Electronic Nano Systems (ENAS),
Chemnitz, Germany

N. Iwamoto et al. (eds.), *Molecular Modeling and Multiscaling Issues for Electronic Material Applications*, DOI: 10.1007/978-1-4614-1728-6_10, © Springer Science+Business Media, LLC 2012

brought forth a procedure to cross-linked structures complying strictly to three-dimensional boundary conditions: the special challenge lies in a cross-linking algorithm that produces cross-links that extend the network beyond the periodic unit cell, interconnecting the latter with its virtual images.

## 10.1 Introduction

In this contribution the potential of molecular dynamics simulation for structure–property correlations in epoxy resins is discussed. This is a topic relevant for a multiscale framework towards lifetime prediction in microelectronics packaging. The need for the development of such a framework arises from an ever-increasing complexity of interaction between materials as dimensions decrease, making efforts necessary to combine numerical methods on continuum mechanics scale (finite element modeling, FEM) down to the molecular (coarse-grained molecular dynamics, CGMD) and even atomic level (molecular dynamics, MD). By utilizing the latter method, where models are built atom by atom representing the exact chemical structure of a material, a deeper understanding can be achieved of the relevant properties that affect a material's performance in the field. While in experiments, synthesis and testing can be time and resource consuming, molecular dynamics simulations offer, once standard procedures have been established, a way to predict some properties and to determine trends in others while changing structural parameters (free volume, chemistry) or load conditions acting on the material (humidity, temperature). The capabilities of molecular dynamics simulations can be exploited in two ways: chemical or physical parameters of materials can be systematically varied at the drawing board and secondly the various interfaces in complex structures can be simulated to give insight into the physics of failure through direct observation of molecular phenomena at the location where failure is likely to occur and experimental data is often hard to obtain.

One of the goals of our research activities in this area was to contribute to laying the basis of experimentally and numerically acquired data in order to establish procedures of atomistic molecular modeling that helps in a comprehensive simulation assisted reliability assessment of materials and material combinations in microelectronic packaging.

Increasing research activity in the field of mechanical and transport properties of epoxy resins in recent years underpins its relevance for microsystems integration: The effect of moisture on polymer properties has been reviewed for packaging-relevant polymers by e.g. [1–6]; others describe the interactions of moisture with the polymer matrix [7–9], especially with respect to diffusion [10–12] versus structure [13, 14].

Molecular modeling of cross-linked structures has been conducted by several authors, e.g. [15–19] developing similar methods to construct atomistic molecular models of cross-linked polymers usually based on commercially important epoxy resins. The investigation of moisture diffusion in polymers or epoxy resins has also attracted interest [20–24] as have the investigation of interfacial or surface properties of epoxies [17, 18, 25, 26]. However, direct comparison to experimental data is scarce and often even literature data is unavailable.

Here we will present an approach of a combined experimental and simulative property analysis of cross-linked epoxy resins. Besides a "quick and dirty" approach for a simple but quick tendency assessment, the research has brought forth a procedure to cross-link structures complying strictly to three-dimensional boundary conditions: the special challenge lies in a cross-linking algorithm that produces cross links that extend the network beyond the periodic unit cell, inter-connecting the latter with its virtual images. A practical procedure of cross-linking will be described that adopts ideas found in the literature [15, 16] but incorporates some own ideas as well, to arrive at sufficiently cross-linked models within reasonable computation time.

The algorithm has been applied to layered structures of epoxy and silicon oxide as well, producing two-dimensional periodic networks in an infinite sandwich layer to investigate adhesion phenomena. This will be discussed in the outlook of this work.

## 10.2  Materials Selection and Characterization

### 10.2.1  Chemical Structure of Epoxy Resins and Structure Property Correlation

Epoxy molding compounds basically consist of two classes of materials, a thermosetting polymer system forming the matrix and the filler particles which are contained within this matrix. The variety of filler particles and their purposes are manifold and will not be the subject of this work. In microelectronics, often silica particles are used and their main purpose is to compensate for the higher coefficient of thermal expansion (CTE) with respect to other materials. The epoxy matrix material is formed from basically two materials, the epoxy resin and a hardener molecule. A wide variety of combinations and different chemistries exist and are applied, but exact formulae are seldom published, making fundamental research difficult on materials that are actually in use. However, common features are known and hence similar materials can be investigated instead.

For our work two different systems have been chosen with the following considerations determining the choice: First, it obviously had to be an epoxy resin which forms a 3D-interlinked network structure. To concentrate purely on bulk properties, an unfilled polymer should be used (no internal interfaces). Second, it had to be a polymer of small molecular weight to have as many epoxy-hardener groups in the model as possible. Third, it had to be an epoxy resin of known composition with the possibility of varying inner variables during in-house

**Fig. 10.1** Chemical formula
of the Epoxy Resin
*1,3-Bis(2,3-epoxypropoxy)-
benzene* (a) and an atomistic
model in ball and stick
representation (**b**).
Coloration: Oxygen: *dark
centers (red* in web version);
carbon: *gray*; hydrogen:
*white*

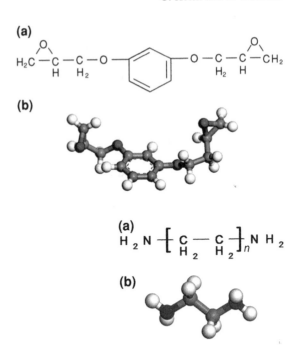

**Fig. 10.2** Chemical formula
of the hardener molecule
*1,2-Diamino(ethane)$_n$* (a) and
an atomistic model in *ball
and stick* representation (**b**).
Coloration: Nitrogen: *dark
centers (blue* in web version);
carbon: *gray*; hydrogen:
*white*

synthesis. This led to the choice of the aromatic epoxy *1,3-Bis(2,3-epoxypropoxy)-
benzene* and a *1,2-diamino(ethane)$_n$* (where $n = 1,2,3$) hardener as depicted in
Figs. 10.1 and 10.2. In the chemical, cross-linking reaction, the epoxy ring will
open at the oxygen–carbon-bond, forming a *hydroxyl*-group at the second carbon
while the first carbon reacts with the nitrogen of the diamine-group of the hardener. While the epoxy is bi-functional, the hardener is tetra-functional, leading in
the polymerization reaction to a 3D-cross-linked structure. By variation of the
stoichiometry of the mixture of the two components from 2:1, where theoretically
all reactive sites can find a reaction partner, to 4:3, more unreacted amino-groups
are expected to increase the free volume and enhance the polarity of these enlarged
sites. By varying the number of ethane groups $n$, the hardener chain is lengthened
to induce structural changes as well (Fig. 10.2).

As detailed above, this combination allows small variations in the chemical
structure of the polymer:

- An alteration in stoichiometry by detuning the epoxy hardener composition from
  a matching 2/1 ratio (theoretically fully reacted) to a 4/3 ratio with then residual,
  un-reacted and highly polar hardener end-groups.
- An alteration in hardener chain length (2, 4 and 6, carbon atoms respectively) to
  vary free volume and overall polarity.

Figure 10.3 sketches the expected changes in structure and free volume by
variation of epoxy/hardener ratio. According to this philosophy, the following test

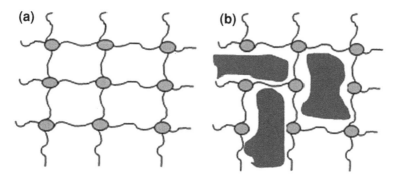

**Fig. 10.3** Sketch of the expected changes in structure and free volume by variation of the epoxy/hardener ratio **a** epoxy/hardener ratio 2/1 **b** epoxy/hardener ratio 4/3 (*Orange circles* denote the hardener molecule (4-functional) while *black lines* represent the bi-functional epoxy)

**Table 10.1** Test matrix: material variables and nomenclature used in this work

| # carbons in diamine chain (2n) | Epoxy/hardener ratio matched | Epoxy/hardener ratio unmatched |
| --- | --- | --- |
| 2 | 2/1-2 | 4/3-2 |
| 4 | 2/1-4 | 4/3-4 |
| 6 | 2/1-6 | 4/3-6 |

matrix can be drawn up (Table 10.1), introducing at the same time the nomenclature for the test polymers.

For the choice of the second epoxy system a more industrial-grade system was aimed for. However, the complex composition of most industrially used molding compounds is a safely guarded secret. Therefore, the model epoxy system Epoxy Phenol Novolac (EPN 1180) was chosen, which is similar to industrially used chip-encapsulation molding compounds [27, 28]. The epoxy resin and the Bisphenol A hardener (*4,4'-dihydroxy-2,2-diphenylpropane*) are depicted in Figs. 10.4 and 10.5.

As can be seen from the chemical formula in Fig. 10.4, the epoxy resin consists of a backbone of aromatic rings connected by methylene groups. The connection can be either para- or ortho-, while the fraction of meta-substitution can be neglected. For the molecular models, the exact distribution was not considered very important, and therefore an arbitrary mix of tri- ($q = 1$) and tetra-functional ($q = 2$) epoxy monomers was used. Using a monomer mixing ratio of tri:tetra:BPA $= 2{:}3{:}9$, the average functionality of $f = 3.6$ was met and the number of BPA hardeners allows a theoretical conversion of 100%.

In the curing reaction, the epoxy ring opens and the carbon atom connects to the oxygen of the BPA hardener, while the hydrogen of the hydroxyl group of the BPA switches to the epoxy-oxygen to form yet another hydroxyl group. At the correct

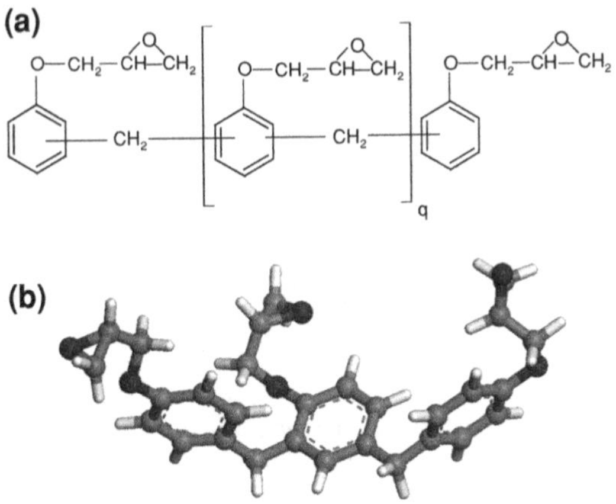

**Fig. 10.4** Chemical formula of the Epoxy Phenol Novolac Resin (**a**) and an atomistic model of a tri-functional molecule ($q = 1$) in *ball* and *stick* representation (**b**). Coloration: Oxygen: *dark centers* (*red* in web version); carbon: *gray*; hydrogen: *white*

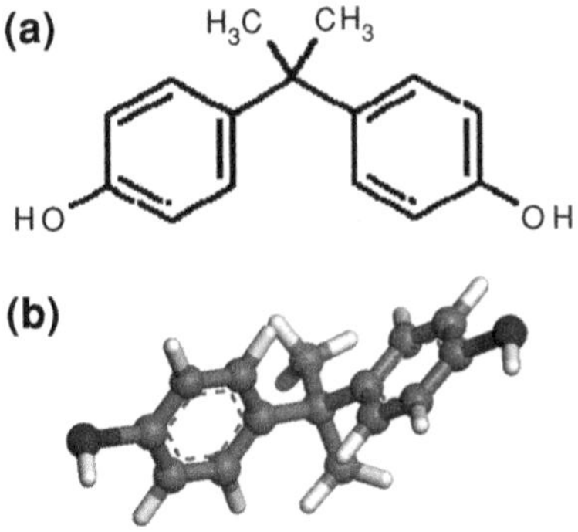

**Fig. 10.5** Chemical formula of the hardener molecule Bisphenol A (**a**) and an atomistic model in *ball* and *stick* representation (**b**). Coloration: Oxygen: *dark centers* (*red* in web version); carbon: *gray*; hydrogen: *white*

curing temperature, this exothermic reaction takes place until either all epoxy groups have found a reaction partner (theoretically 100% conversion) or until, in the course of the glass transition, the mobility of reaction partners has decreased so far that further reaction is practically impossible. A typical test for the degree of conversion by DSC or other means will show that full conversion has been achieved (technically 100% conversion); however, the absolute degree of conversion remains unknown. This uncertainty has to be kept in mind when comparing experimental and simulated properties.

### 10.2.2 Material Characterization

One guiding idea during this work was to have simple and fast standard characterization methods of known accuracy and reproducibility at hand. This will make it easier to interpret the results and extend the method to industry grade polymers

**Table 10.2** Overview characterization Methods

| Investigation | Experiment | Conditions | Used on epoxy system |
|---|---|---|---|
| Density ($\rho$ [g/cm$^3$]) | Dilatometry | T = 25°C | 2/1-(2n), 4/3-(2n), EPN |
| Thermal expansion (CTE, [ppm/K]) | TMA | T = 25–180°C | 2/1-(2n), 4/3-(2n), EPN |
| Storage modulus (E [MPa]) | DMA (1 Hz) | T = -40 to 140°C dry and wet | 2/1-(2n), 4/3-(2n), EPN |
| Glass transition temperature ($T_g$, [°C]) | DMA (1 Hz) | T = -40 to 140°C dry and wet | 2/1-(2n), 4/3-(2n), EPN |
| Hygroscopic swelling ($\varepsilon_\mu$, [%]) | TMA, Desorption | T = 105°C | 2/1-(2n), 4/3-(2n) |
| Diffusion coefficient (D [cm$^2$/s]) | Gravimetric; ab- and de-sorption | See Table 10.3; H-const, T-vary | 2/1-(2n), 4/3-(2n), EPN |
| Water uptake ($C_{sat}$ [%wt]) | Gravimetric/ Absorption | See Table 10.3 | 2/1-(2n), 4/3-(2n), EPN |

**Table 10.3** Test matrix: External load variables for materials 2/1-(2n) and 4/3-(2n)

| | T [°C] | H [%r.h.] |
|---|---|---|
| Level 1 | 65 | 65 |
| Level 2 | 85 | 85 |
| Level 3 | 100 | 98 |

later. The test program put into effect during the preliminary study is depicted in Table 10.2. It must be noted that the test-program for the second material EPN was not as extensive since it was not synthesized in house, and therefore some of the properties were taken from the literature or data-sheets.

### 10.2.2.1 Density

A very important parameter is density $\rho$(T,C) as function of temperature T and moisture concentration C(T) in the polymer, as it is needed for comparison with the equilibrated MD model. Density for the 2/1-(2n) and the 4/3-(2n) materials was measured using a mercury dilatometer at room temperature. The density at room temperature (RT) is given in Table 10.4 before and after curing (cure shrinkage is around 9% for all groups). As can be seen, there is a monotonous decrease in density with increasing number of carbon atoms in the hardener chain, increasing the volume more than the mass of the polymer in all cases. In general the density of the unmatched system (4/3-(2n)) is slightly lower. This is due to a proportional over-representation of lighter hardener molecules and a lower cross-linking density after curing.

**Table 10.4** Experimental results for density and CTE

| Material | Density | Density | $T_g$ | CTE | CTE |
|---|---|---|---|---|---|
| 11 | Prior-shrink | Post shrink | Dry state | $T < T_g$ | $T > T_g$ |
| 12 | g/cm$^3$ | g/cm$^3$ | °C | ppm/K | ppm/K |
| 2/1-2 | 1.16 | 1.27 | 94 | 52 | 174 |
| 2/1-4 | 1.14 | 1.24 | 82 | 50 | 175 |
| 2/1-6 | 1.12 | 1.21 | 75 | 55 | 183 |
| 4/3-2 | 1.14 | 1.26 | 72 | 53 | 190 |
| 4/3-4 | 1.11 | 1.22 | 62 | 54 | 201 |
| 4/3-6 | 1.09 | 1.18 | 58 | 64 | 207 |
| EPN | – | 1.18* | 110 | 193 | 570 |

* Literature value [28]

**Fig. 10.6** TMA measurement on EPN showing the change in CTE at the glass transition

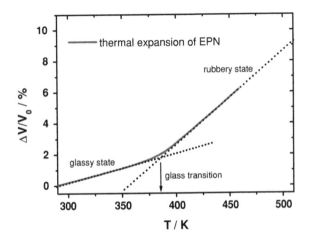

### 10.2.2.2 Thermal Expansion

The coefficient of thermal expansion (CTE), $\alpha(T)$, was measured with TMA (thermo-mechanical analysis). The correlation to the thermal strain $\varepsilon_\vartheta(T)$ is given by

$$\varepsilon_\vartheta(T) = \int_{T_0}^{T} dT\, \alpha(T) \tag{10.1}$$

A typical curve for the EPN sample is shown in Fig. 10.6 and values are given in Table 10.4. For the 2/1-(2n) and the 4/3-(2n) materials there is a clear tendency, although the given values are averages for the specific T-intervals: The more carbon atoms in the chain, that is, increasing $n$, the higher the CTE will be. This is true below and above the glass transition temperature $T_g$ for all materials, whereas the unmatched configurations tend to have higher CTE. Here, $T_g$ is taken at the kink in the $\varepsilon_\vartheta(T)$ curve.

**Fig. 10.7** DMA
measurement on EPN
showing a drop in storage
modulus and a peak in phase
angle tan($\delta$) at the glass
transition

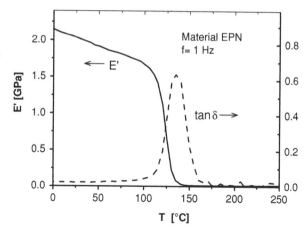

### 10.2.2.3 Storage Modulus and Glass Transition Under Moisture

The storage modulus E and glass transition temperature $T_g$ as influenced by
moisture were characterized by dynamical mechanical analysis (DMA) at a frequency of 1 Hz, shown for the EPN material in Fig. 10.7. $T_g$ was determined by
the maximum of the phase angle tan($\delta$) and is as usual higher than the TMA value
(given in Table 10.4 for the dry state). For the 2/1-2 material DMA measurements
at different moisture loading conditions are shown in Fig. 10.8. Whereas CTE and
density are rather typical for an unfilled epoxy-resin, $T_g$ is rather low. As seen in
Fig. 10.8, this $T_g$ drops to even lower values in the presence of moisture loading H.

The water molecules seem to weaken the intermolecular forces for the polymer
to prefer the state of entropy elasticity already at lower temperature. This
dependence of $T_g$ on moisture content is for all variants given in Fig. 10.9. Very
conspicuous are the unmatched groups (4/3-(2$n$)) as here $T_g$ drops to near room
temperature (RT).

The effect of $T_g$ drop has repercussions on the storage modulus. Accordingly, the
modulus also drops to much lower values to the effect that for the unmatched groups
near RT there is hardly any stiffness left (see Fig. 10.10). As moisture and temperature within a polymer are not independent the specimens have been measured in a
moist environment during DMA testing below T = 100°C. It has been found that this
was not necessary as moisture concentration change by desorption happens on a
much larger timescale as the temperature variation in the DMA.

### 10.2.2.4 Moisture Uptake and Diffusion Coefficient

Maximum moisture concentration $C_{sat}(T)$ for a certain temperature and external
moisture concentration $H$ was measured by gravimetric measurement. Therefore a
flat sample with area 50 × 30 mm$^2$ and a thickness of $h$ = 2 mm was used for a

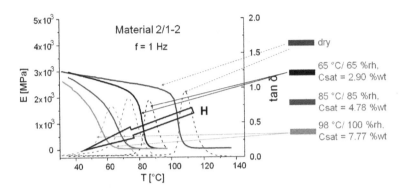

**Fig. 10.8** $T_g$ drop with increasing moisture content as measured by DMA

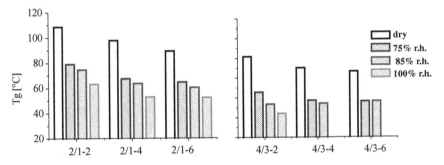

**Fig. 10.9** $T_g$ by DMA as a function of loading variables $C$ for the 2/1 and 4/3 materials. For the 4/3-4 and 4/3-6 materials at the highest moisture content $T_g$ drops below RT (Vertical axes are the same)

one-dimensional mass transport approximation that allowed diffusion coefficient $D(T, H)$ evaluation at the same time. Using the time-honored method of scaling the x-axis by $t^{0.5}$, one may calculate the diffusion coefficient $D$ by using the initial slope of the absorption curve governed by

$$\frac{m(t)}{m_\infty} \approx 4\sqrt{\frac{Dt}{\pi h^2}} \tag{10.2}$$

where m($t$) is the mass of the absorbed moisture in the polymer and $m_\infty$ its value at saturation. This formula can be derived when solving the partial differential equations of diffusion (i.e. Fick's second law) analytically by a Fourier series approach (see e.g. [29]) and approximating for small $t$. Such a representative curve depicted in Fig. 10.11, showing $C(t) = m(t)/m_p$ with the dry polymer mass $m_p$.

For all epoxy resins purely Fickian diffusion was observed and all samples have been loaded to reach saturation mass. All values given in this paper refer always to fully saturated samples with $C_{sat} = m_\infty/m_p$.

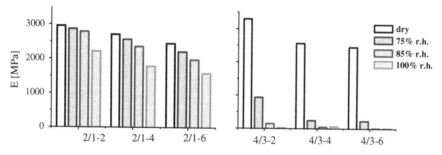

**Fig. 10.10** Storage modulus at T = 35°C as function of moisture loading H. Samples were always saturated (Vertical axes are the same)

**Fig. 10.11** For all materials purely Fickian diffusion was observed. Note the very good fit by analytical or FE-simulation result and that EPN shows much less saturation concentration than 2/1-2

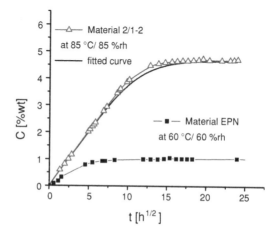

The results of the moisture uptake at saturation concentration $C_{sat}(T)$ are given in Fig. 10.12 for the 2/1-($2n$) and 4/3-($2n$) materials. First of all, the very high moisture concentration is noteworthy. Compared to highly filled epoxy systems often used in electronics and display $C_{sat} \approx 0.1\ \%$ (see e.g. [5] and EPN in Fig. 10.11) these polymers act like a sponge. $C_{sat}$ is observed to increase with simultaneous moisture and temperature loading. As given e.g. in [2], $C_{sat}$ is only expected to depend on H below $T_g$ for polymers that do not chemically react with water. Above $T_g$, however, $C_{sat}$ may increase with T due to an increase in free volume [12]. Further, it can be observed that with increasing number of carbon atoms in the chain now there is less water that can be stored in the matrix. So although there is a lower density and hence more free volume, less water can be taken up. This might be due to the fact that with increasing chain length the overall polarity decreases. This is supported by the fact that hydrophilic (polar) groups can congregate water by hydrogen bonding [9]. This behavior is observed for both matched and unmatched groups. For the unmatched groups providing excess polarity due to remaining end-groups and more free volume $C_{sat}$ is consequently higher [10].

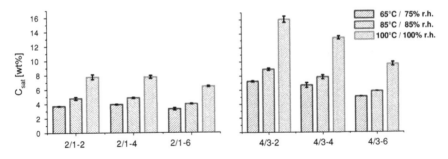

**Fig. 10.12** Saturation concentration as function of loading conditions for all polymer variants

**Fig. 10.13** Diffusion coefficient function of loading conditions for all polymer variants. So far, $H$ and $T$ are varied simultaneously

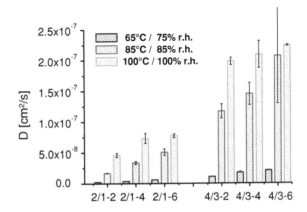

The results from diffusion coefficient measurement are depicted in Fig. 10.13. Here, one observes a very distinctly resolved dependence with values spanning more than one order of magnitude within the given testing conditions. Apparently moisture transport through the matrix is very sensitive to the internal variables of Table 10.1 and the loading conditions in Table 10.3.

From Fig. 10.13 one infers that $D$ increases with decreasing density (i.e. increasing free volume) and is also significantly higher for the unmatched groups. The latter might be due to a two-phase microstructure which has been found for then non-stoichiometric resins [13] showing a low-density phase. With respect to temperature one obviously expects an exponential behavior, since it is well known that the diffusion coefficient $D(T)$ obeys an Arrhenius-type law with activation energy $E$ and Boltzmann constant k:

$$D \propto \exp\left(-\frac{E}{kT}\right) \tag{10.3}$$

In a diagram of $\log(D)$ against $T^{-1}$ the expected result according to Eq. 10.3 is a straight line. For that purpose samples of identical $C_{sat}$ [conditioned at (85/85)] were desorbed at different $T$ to allow a cross correlation between temperature and

**Fig. 10.14** Diffusion coefficient as function of temperature according to Eq. 10.3. Note that some materials have their $T_g$ within the tested interval

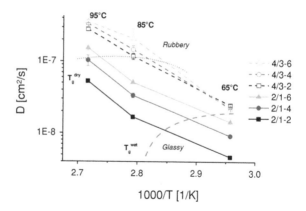

moisture. For all materials this is depicted in Fig. 10.14, where the $T_g$ in the saturated state is depicted as a dashed line. As can be seen, there are deviations from straight lines which could be due to the fact that some materials cross from the rubbery into the glassy state upon desorption, thus obeying two different activation energies [11, 14]. To within measurement accuracy, only the unmatched groups, which are nearly always in the rubbery state, show behavior close to a straight line.

### 10.2.2.5 Moisture Swelling

Moisture swelling was measured with TMA desorbing fully saturated samples of different $C_{sat}$ at T = 105°C (slightly higher than all other temperatures used). Upon plotting the hygroscopic stain $\varepsilon_\mu$ versus moisture concentration within the sample one gets a linear relationship as shown in Fig. 10.15. Also the relationship $C_{sat}$ versus $H$ is approximately a linear one at constant temperature (not shown here) [13].

This has also been found by e.g. [5 or 2]. Apparently all moisture contributes to swelling strain [7, 8], either by filling up the free volume (unbound water) or by interfering with the intermolecular potentials between the chains (bound water), causing weaker hydrogen bonds between thus expanding the volume [6]. The result given in Fig. 10.15 allows to calculate a coefficient of moisture expansion (CME) to approximately $\beta \approx 0.405$ [ppm/%] common to all the six polymers to within measurement accuracy. This suggests that the mechanism behind the expansion should be the same for all, too.

### 10.2.2.6 Free Volume Investigation

As the presented results are well known in the literature, sorption, diffusion and other mechanical properties of polymers seem to be connected to their free volume, since it provides space for penetrating molecules and influences the

**Fig. 10.15** Moisture swelling as function of moisture concentration. The correlation is linear and nearly the same for all polymers under test and not explicitly dependent on temperature. A small kink for 2/1-2 might be due to its $T_g = 94°C$ close to the testing $T = 105°C$

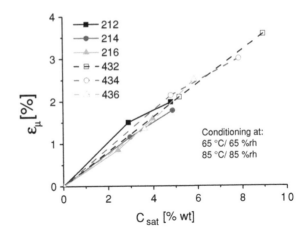

mobility also for chain segments of the polymer matrix. However, experimental methods to determine the free volume are scarce, and there exist multiple definitions of the free volume. Here, we have always focused on the accessible free volume which may be obtained by the experimental method of PALS (positronium annihilation lifetime spectroscopy). In the PALS method, the lifetime of a positronium can be related to the average hole size of the free volume distribution. In this technique, positronium in its ortho-state, o-Ps, is used as a probe for local free volumes. Assuming spherical holes and further simplifications, the hole size can be calculated from the o-Ps lifetime $\tau_3$ using a quantum mechanical Tao-Eldrup model with an empirical parameter. Details about the experimental setup and calculation methods can be found in [30]. It should be noted that from the PALS measurements, an average hole size of the free volume distribution is gained but not the absolute amount of free volume, since the method does not yield a measure of the number-density of holes. Furthermore it should be noted that hole geometries that deviate from a spherical shape may lead to an underestimation of its volume (see Fig. 10.17). By analyzing the o-ps lifetime $\tau_3$ at different temperatures (see Fig. 10.16), the thermal expansion of the epoxy systems manifest in a rise in average hole-volume.

The glass transition temperature $T_g$ is marked by a distinct change in slope of the $\tau_3(T)$ curve, following the change in CTE. The values for $T_g$ from PALS and TMA are in acceptable agreement for the systems with short hardener chains (E/H-2) (see Table 10.5), while for the E/H-6 systems a large deviation and no clear trend is observed. Further measurements will have to clarify the cause of this variance which may well root in a deviation of the hole geometry from the assumed spherical shape that could account for decreasing $\tau_3$ as is sketched in the sample preparation. From the values of $\tau_3$ at $T = 30°C$, the average hole volumes of a free volume distribution were calculated for four of the systems. In Table 10.5 these values are listed.

**Fig. 10.16** Average hole free volume can be calculated from positron lifetime $\tau_3$ measured by PALS

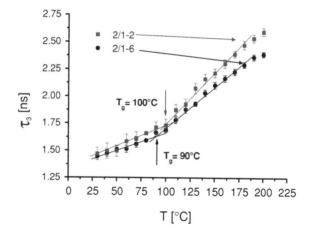

**Fig. 10.17** Possible deviations of hole geometry from the assumed spherical shape that could lead to detection of decreased $\tau_3$ values

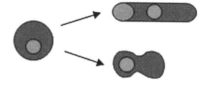

**Table 10.5** Free volume data

| Material | $\Phi$ [%] | $T_g^{TMA}$ [°C] | $T_g^{PALS}$ [°C] | $\tau_3$ [ns] | $V_{hole,avg}$ [Å$^3$] |
|---|---|---|---|---|---|
| 2/1-2 | 10.01 | 94 | 100 | 1.469 | 53.2 |
| 2/1-4 | 10.67 | | | | |
| 2/1-6 | 11.54 | 75 | 90 | 1.449 | 51.6 |
| 4/3-2 | 10.67 | 72 | 80 | 1.399 | 47.3 |
| 4/3-4 | 11.45 | | | | |
| 4/3-6 | 12.67 | 58 | 40 | 1.505 | 56.5 |

## 10.2.3 First Conclusions

From these preliminary experiments it is clear that the chosen internal material parameters affect the macroscopic behavior of the polymers. The diffusion coefficient D has been identified to be most sensitive to the internal and external variables and therefore is best suited for establishing a structure–property correlation by MD simulation. So far, free volume and polarity seem, as inferred from the chemical structure, to be the main determinants with respect to the polymer structure, as can be qualitatively seen in first experiments. Detailed atomistic molecular modeling can help understand these results on the molecular level.

## 10.3 Molecular Modeling of Cross-linked Epoxy Structures

### 10.3.1 Force Field, Scale Considerations and Periodic Boundary Conditions

In molecular dynamics simulations, an ensemble of atoms is described by the individual positions $\mathbf{r}_i$ within a simulation cell, the bond information between the atoms and a vector $\mathbf{v}_i$ describing the current velocity of each atom. The dynamics of the system may now be determined using the force field information which describes the interactions of bonded atoms as sketched in Fig. 10.18 (bond-lengths, bond-angles, conformation) and non-bonded interactions (van der Waals, electrostatic).

Force fields thus allow the calculation of the potential energy $E$ of an ensemble of $N$ atoms, as a function of their coordinates $(\mathbf{r}_1 \ldots \mathbf{r}_N)$:

$$
\begin{aligned}
E(\mathbf{r}_1 \ldots \mathbf{r}_N) = & \sum_{bonds} \text{bond - length - deformation} \\
& + \sum_{angles} \text{bond - angle - deformation} \\
& + \sum_{conf.angles} \text{torsional - deformation} \\
& + \sum_{atom-pairs} \text{nonbonded - interactions}
\end{aligned}
\tag{10.4}
$$

For a more detailed description of the interaction contributions described by a force field see e.g. [21, 31, 32]. Throughout this work, the commercial force field COMPASS of Accelrys Inc. is used [33]. For a given molecular structure, the force field results in a potential energy surface. Integration of the Newtonian equations of motion leads to a new velocity of each particle which can be extrapolated over the time step of the simulation to determine the new coordinates $\mathbf{r}_i$.

$$
\mathbf{F}_i = -\nabla_{\mathbf{r}_i} E_i(\mathbf{r}_1 \ldots \mathbf{r}_N) = m_i \ddot{\mathbf{r}}_i
\tag{10.5}
$$

The force $\mathbf{F}_i$, acting on a particle $i$ of mass $m_i$, results from the gradient of the potential energy $E_i$ (Eq. 10.5) determined by the force field (Eq. 10.4). As time step for a simulation, a value must be chosen that ensures accurate representation of the fastest vibration in the system (i.e. C–H bond at a period of $\tau \approx 10$ fs), which usually leads to the choice of $\Delta t = 1$ fs. Knowing that in order to achieve reliable results, several hundred picoseconds of molecular dynamics simulation should be conducted, the vast computational effort for large systems becomes obvious. However, even large models of several thousand atoms result in volumes of the order 50 nm$^3$ and therefore do not adequately represent bulk behavior, since the surface-to-volume-ratio is by far too large. To still be able to predict bulk properties, three-dimensional periodic boundary conditions (3D-PBC) are usually

bond length                     bond angle                          torsion

**Fig. 10.18** Sketch of possible deformations covered by a force field in molecular dynamics simulations

applied. While 3D-PBCs help in reducing the computational effort, they also introduce the risk of

- replicating artifacts and thereby multiplying their relevance
- periodicity of nonbond-interactions (special algorithms or cutoffs usually deal with that)
- nanograin-simulation (see Fig. 10.29 and discussion)

However, these risks can be well taken care of if the dimensions of the model are reasonably large, models are well prepared, checked for possible artifacts and all technical simulation details are set accordingly.

### 10.3.1.1  Branched Approach

The cross-linking reaction as it is observed in the experiment is a complex quantum-mechanical process involving the breakage of bonds, hydrogen transfer and the formation of bonds. Since the investigation of this process by detailed atomistic simulations is beyond the scope of these works, the challenge was to develop a convenient procedure of simulating epoxy resins. Two approaches have been followed, the first one being a method to avoid actual cross-linking while retaining chemical and the most important structural characteristics (besides 3D-cross-links). The second, emanating from the first, is the development of a cross-linking algorithm which is practical and fast but still near enough to reality to lead to packing models that represent bulk behavior of the epoxy network. The latter will be discussed further below to show and discuss the results of the "quick and easy" approach first.

To model the epoxy resin without cross-links, essentially means treating it as a (viscous) liquid. If this was done with an unit cell filled with the molecules depicted in Fig. 10.1, it would not be realistic as highly polar end-groups are still present which vanish during the chemical reaction. Here, it was tried to eliminate these groups ($-NH_2$ and epoxy groups) by interlinking both ends of two epoxies with a single hardener (forming four bonds), then opening each epoxy ring structure at the third carbon atom. All remaining open ends are then saturated with hydrogen. Thus the overall molecular weight (save two hydrogen atoms per severed bond) and the chemical characteristics of a reacted $E–H$ system (like e.g. the nitrogen and oxygen environment) are conserved. This simplification is depicted in Figs. 10.19 and 10.20. This structure was chosen from a set of different possible structures as it showed the lowest polarity, i.e. the best saturation of reactive end-groups.

**(a)**

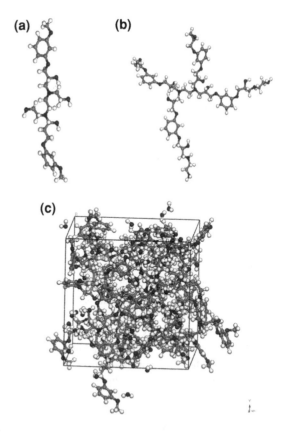

**(b)**

**Fig. 10.19** Simplified chemical structure used to mimic a 3D cross-linked network by a liquid using partially reacted molecules of specified stoichiometry **a** 2/1-2 **b** 4/3-2

**Fig. 10.20** **a** The sub-entities are depicted for a matched system (2/1-2) **b** an unmatched system (4/3-2) and **c** Unit cell with periodic boundary conditions for bulk diffusion calculation. (Here, the solid is effectively modeled as a viscous liquid as no 3D cross-linking is present yet)

**(a)**          **(b)**

**(c)**

Models to be investigated were thus prepared by packing a number of the branched molecules (Fig. 10.18) into a unit cell, applying 3D-PBC and choosing cell size according to the measured density information from experiment (Table 10.4).

## 10.3.1.2 Moisture Transport and Mobility

To investigate transport phenomena in molecular dynamics simulations, often the mean-squared-displacement $s(t-t_0)$ (sometimes: MSD) of a set of $N$ atoms is analyzed (e.g. that of the oxygen atoms of the water molecules):

$$s(t - t_0) = \frac{1}{6N} \sum_{i=1}^{N} \left\langle [r_i(t) - r_i(t_0)]^2 \right\rangle \tag{10.6}$$

where $r_i(t)$ denotes the position of the atom at time $t$ and the angled brackets denote time averaging over all possible choices of time origin $t_0$. In the course of an MD-simulation, the movement of one of these atoms, observed from one structural "snapshot" to the next, will consist of smaller displacements within an energetically favorable "site" of a more vibrational nature and it will consist of diffusive jumps from one of these sites to the next. These diffusive jumps lead over some time to a random walk of the molecules through the epoxy matrix. For Fickian diffusion in the limit of large times, the diffusion coefficient $D$ can be obtained directly from the mean square displacement [21, 23]:

$$D = \lim_{t \to \infty} \frac{d}{dt} s(t - t_0) \tag{10.7}$$

The condition of large times can be checked in a log–log plot of the MSD versus time, where the slope of the linear curve should be unity, that is, the relationship between MSD and time is linear.

A closer look at Eq. 10.6 reveals the role of diffusion-related jumps. In a cubic lattice of grid spacing $a$, the MSD can be expressed with the help of the Einstein relation [34]:

$$s(t) = na^2 = 6Dt \quad \rightarrow \quad n = 6Dt/a^2 \tag{10.8}$$

where $n$ is the number of jumps and $a$ is the mean jump distance of diffusion jumps. Equation 10.8 provides the means to estimate how often jump-events can be expected to occur in a simulation of a given time $t$, provided the diffusion coefficient $D$ is known and a reasonable estimate of the jump distance is assumed. The latter can be provided by an analysis of a simulation run: Fig. 10.21 shows a plot of the distances $r$ of 10 water molecules from the origin of the simulation cell.

In the course of a two-nanosecond simulation, it can be observed that while some molecules merely perform the "vibrational" motion as discussed above, others cover a larger distance which can be assumed to contribute to diffusion. The jump distance $a$ can be roughly estimated to be $a \approx 5$ Å in this analysis of the EPN system. It is worth to note that at this estimated jump distance, the squared displacement after 1 jump amounts to 25 Å$^2$. Plots of the MSD can thus be checked for plausibility [35].

An analysis of the mean squared displacement is shown in Fig. 10.22 for the system of 2/1-2. Into eight independent structural configurations, a number of

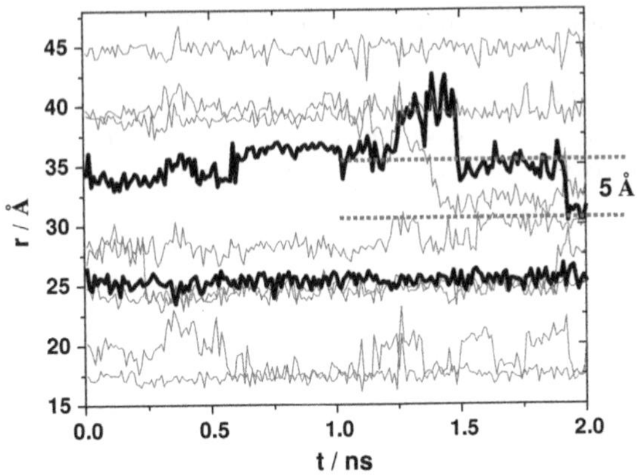

**Fig. 10.21** Distances $r$ of
$H_2O$ molecules from the
origin of the simulation cell.
*Bold lines* depict one example
of motion including diffusive
jumps and another one with
vibrational motion only.
Estimated jump distance of
5 Å is indicated by *dotted
lines*. Note that this is an
analysis in the Epoxy Phenol
Novolac system

**Fig. 10.22** The mean square
displacement $s(t)$ for
calculation of $D$. The scatter
is due to the finite size of the
unit cell and eight different
structural initial conditions

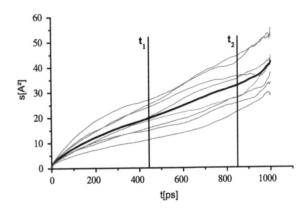

$N$ water molecules have been inserted into the matrix corresponding to a set of
thermal and moisture conditions. The (time-averaged) diffusion coefficient can be
obtained by using Eq. 10.7. Although defined in the limes, for practical reasons
$D$ can only be calculated averaged over a finite interval in time $t \in [t_1; t_2]$ as
indicated in Fig. 10.22: Here, the average of the slopes on curves of eight different
initial configurations has been calculated using the interval where there is a linear
slope (indication of net water movement by hopping mechanism) [21, 23, 36].

The first and last part of the curves must be neglected: For $t < t_1$, the limes
requirement is obviously not valid and equilibration still going on, for $t > t_2$ not
enough $\Delta t$ intervals can contribute any more to give reasonable statistics. For all
simulations the averaging interval was chosen from $t_1 = 450$ ps to $t_2 = 850$ ps for
best linearity.

It must be noted that the plausibility check mentioned above, leading to a
displacement $d = s(t)^{0.5} \approx 5.5$ Å, suggests that these results should be treated
with care. However, the jump distance for the 2/1-2 material is expected to be
lower than that for EPN shown in Fig. 10.21 because smaller chemical compounds

**Fig. 10.23** D as function of density $\rho$. Test runs ($n = 3$) per 2/1-2 configuration in NVT: $t = 1$ ns, unit cell size 22.4 Å, $C_{sat} = 4.5\%$wt corresponds to 16 E–H-E groups and 23 $H_2O$

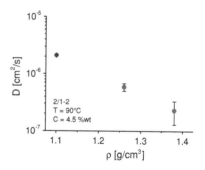

are involved. A thorough analysis on jump distances in 2/1-($2n$) and 4/3-($2n$) materials has yet to be done.

### 10.3.1.3 Sensitivity Analysis

For realistic CPU times the unit cell with its finite size is only an approximation to the real material. Therefore one always conducts a numerical experiment, necessitating multiple runs thus obtaining a mean value and a standard deviation. To determine a minimum unit cell size that balances accuracy to CPU time, multiple runs with different sizes were conducted maintaining the water to epoxy resin ratio. For small unit cell sizes (i.e. $1 < 18$ Å) no linear regime for $s$ was found. So a minimum cell size of $l_{min} \approx 23$ Å and a number of acceptable (i.e. linear slope for $s$-curve) configurations. The influence of the important parameter of density is given in Fig. 10.23 as an example. The uncertainty increases with density for the same computation time (here $t = 1$ ns) as hopping events occur less often. $n_{min} = 4$ could be specified to obtain a diffusion coefficient to within $\sigma \approx \pm 20\%$ accuracy.

### 10.3.1.4 Density Calculations

As was determined in the paragraph above, density is an important material parameter and properties of interest like the diffusion coefficient may be very sensitive to it. For MD simulation one needs the density of the material either as boundary condition for NVT ensembles or as corrective parameter in NpT ensemble calculations. From the density measurements at RT in Table 10.4 we can calculate the density as a function of temperature for the dry state by simply taking into account volume expansion: $\rho(T) = \rho(RT, 0)$

$$\rho(T) = \rho(RT, 0)\frac{1}{1 + 3\varepsilon_v} \tag{10.9}$$

This can be done in a straightforward way using the TMA measured $\varepsilon_9(T)$ values. The resulting densities are given in Table 10.6. The case with moisture is

**Table 10.6** Density evaluation data

| Density [g/cm³] | 2/1-2 | 2/1-4 | 2/1-6 | 4/3-2 | 4/3-4 | 4/3-6 | Method |
|---|---|---|---|---|---|---|---|
| $\rho$ (25°, dry) | 1,27 | 1,24 | 1,21 | 1,26 | 1,22 | 1,18 | Measured |
| $\rho$ (100°C, dry) | 1,25 | 1,21 | 1,18 | 1,23 | 1,18 | 1,14 | Calculated |
| $\rho$ (100°C, 100%rh) | 1,24 | 1,20 | 1,17 | 1,20 | 1,16 | 1,13 | Extrapolated |

The data in the last row was taken for MD-simulation input under NVT

not so simple: As it was not measured yet, we can only calculate a lower threshold for $\rho(T, H)$, assuming that all moisture uptake is completely converted into hygroscopic strain $\varepsilon_\mu(T, H)$ to yield:

$$\rho(T) = \rho(RT, 0)\frac{1 + C_{sat}(T, H)}{1 + 3\varepsilon_v + 3\varepsilon_\mu + 9\varepsilon_v\varepsilon_\mu} \tag{10.10}$$

According to Fan [2], both hydrogen bonded water in the matrix as well as non-bonded water in the free volume contribute to swelling, the latter to a less significant part. Figures 10.12 and 10.15 suggest that these mechanisms occur simultaneously and we can extrapolate linearly to obtain the total volume for the configuration (100, 100) by $\varepsilon_\mu(T, H) = \beta\, C_{sat}(T, H)$. Using Eq. 10.10, values for $\rho(100, 100)$ have been obtained accordingly and are given in Table 10.6. They are not very different from the dry ones, i.e. the excess weight nearly compensates the volume expansion due to moisture swelling.

Based on the density values compiled in Table 10.6, and on the experimental moisture saturation data (Fig. 10.12), saturated packing models were created and a mean square displacement analysis performed for all configurations at 100°C/ 100%r.h. The results are shown in Fig. 10.24 in comparison with experimental data on a logarithmic scale. It can be seen that simulations, although absolute values differ by approximately one magnitude, show good qualitative agreement and represent the trend of higher diffusion coefficients with longer chains (2n) and variation of epoxy hardener ratio (E/H).

In a preliminary study, the effect of cross-linking on the results of the diffusion coefficient was studied. The details of the cross-linking procedure are discussed in the paragraph on cross-linked models. As seen in Fig. 10.24, the non-crosslinked and the cross-linked model ($\approx 98\%$ conversion) are compared to the experiment. As supposed after the results in Fig. 10.24, the results of the more realistic cross-linked model are much closer to the experiment at the same density, obviously modeling the chain mobility more adequately.

### 10.3.1.5 Free Volume Analysis

To compare the results of a free volume analysis with experimental PALS data (see Sect. 10.2.2.6 and Fig. 10.16), we have focused our efforts besides the determination of the total free volume by the method of Bondi [37], on the

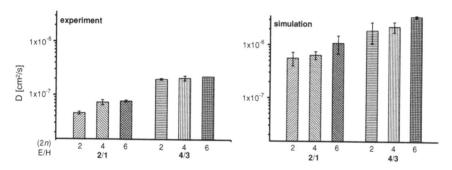

**Fig. 10.24** Diffusion coefficient measured and calculated (no 3D cross-linking) for the systems at condition 100°C/100%r.h. A qualitative agreement can be reached

accessible free volume which is based on the insertion of a test particle [38]. To estimate the total amount of free volume in a polymer (at $T = 0$ K), the method of Bondi can be employed, where the free volume $\Phi$ is defined as the space within the polymer which is not occupied by atoms of the polymer matrix:

$$\Phi = 1 - 1.3 \frac{V_{vdW}}{V_s} \tag{10.11}$$

where $V_s = \rho^{-1}$ is the specific volume of the polymer and the universal packing coefficient, equal to 1.3, is used to convert the van der Waals volume of the repeat unit into the occupied volume. The (specific) volume $V_{vdW}$ of the repeat unit was calculated by the computer program *Synthia* of Accelrys Inc. [39]. Values thus obtained are listed in Table 10.5 in Sect. 10.2.2.6. It can be observed that within each group (2/1 and 4/3) the free volume increases with the chain length of the diamine and that at same hardener chain length, the 4/3 group contains a higher free volume than the 2/1 group, as was expected and intended.

Insight about the structure of the free volume on a molecular level may also be gained by directly probing the packing models by a test particle of similar size as assumed for o-Ps and the distribution of hole sizes may be analyzed, as this method can be compared to PALS. See [38, 40–42] for a detailed discussion. Figure 10.25 shows preliminary results for an investigation of a cross-linked 2/1-2 packing model analyzed directly after construction, the same model after an NVT- and NpT-equilibration and experimental PALS result. As is common, the distribution of fractional free volume (FFV) is plotted against the radius of a volume-equivalent sphere (r.e.s.). The distribution in Fig. 10.25 therefore gives information about a size distribution of holes. Noteworthy is that after construction a large fraction of free volume is concentrated in rather large connected free volume sites that vanish upon equilibration. After equilibration, the distribution peaks at values slightly lower than experimentally determined by PALS, which may be due to a higher cross-linking degree in the model ($\sim 88\%$). Further studies need to be carried out for confirmation and clarification of results.

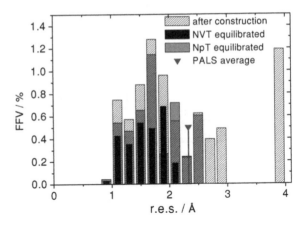

**Fig. 10.25** Free volume distribution in an atomistic model of 2/1-2 by test-particle insertion and experimental PALS result

### 10.3.1.6 Conclusion of Branched Approach

In this section we described an approach and its results to build packing models of the epoxy-systems that avoids chemical cross-links for computational purposes. Chemically identical to real systems, the fragments were of a more branched nature and it was expected that by mechanical cross-linking (interlocking) the system would be represented fairly well. The approach has the advantage that construction of these models is quick and the possibility to produce artifacts (such as the ones described below) is small. Furthermore, it is possible to automatically define charge groups which are electrically neutral to the outside (e.g. $-CH_3$) and for which the long ranging Coulomb interaction does not need to be calculated in full detail. This reduces preparation and computational effort considerably. However, while still being a worthwhile approach, it must be acknowledged that details of the structure of such models may not be represented well enough for certain analyses. So the behavior of the models under NpT equilibration (fluctuating volume) or the distribution of free volume is expected to be considerably influenced, altering the diffusion coefficient. It was therefore attempted to develop a cross-linking algorithm that generates packing models containing chemical cross-links that also obey 3D-PBCs.

## 10.3.2 The 3D Cross-linking Approach

### 10.3.2.1 Cross-linking Algorithm

The cross-linking reaction as it is observed in the experiment is a complex quantum-mechanical process involving the breakage of bonds, Hydrogen transfer and the formation of bonds. Since the investigation of this process by detailed atomistic simulations is beyond the scope of this work, the challenge was to develop a cross-linking algorithm which is practical and fast but still near enough

**Fig. 10.26** Cross-linking
yields more realistic values
for $\underline{D}$. Test runs ($n = 1$) per
4/3-6 configuration in NVT:
$\underline{t} = 1.5$ ns with $t_1 = 0.8$ ns
and $t_2 = 1.3$ ns, dimensions
of cubic unit cell $a = 25$ Å

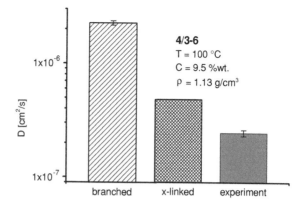

**Fig. 10.27** The cross-linking
algorithm determines if the
reaction sites of opposite
charges (BPA-oxygen: *left*
($-$) charge), epoxy-carbon:
*right* ($+$) charge) are in close
contact

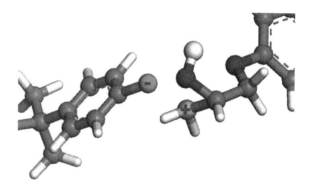

to reality to lead to packing models that represent bulk behavior of the epoxy network. On the basis of the algorithm first tested on the 4/3-6 epoxy system for comparison to branched models (see Fig. 10.26), the procedure was adapted to generate cross-linked models of Epoxy Phenol Novolac (EPN), using the epoxy-hardener system depicted in Figs. 10.4 and 10.5.

Before packing a mixture (see stoichiometry details in Sect. 10.2.1) of epoxy and BPA monomers at target density of the cross-linked system in a simulation cell, with three-dimensional periodic boundary conditions (3D-PBC) applied, the epoxy rings are opened and the hydrogen transfer from the BPA-hydroxyl group to the epoxy-oxygen is performed, leaving a carbon (epoxy) and an oxygen (BPA) as reactive sites. These reactive sites are rigged with opposite coulombic charges of $+1$ and $-1$. This preparation is shown in Fig. 10.27. The monomer-mixture is then subjected to a minimization—MD-simulation-cross-linking loop:

1. Geometry optimization of the structure
2. Molecular dynamics simulation
3. Close contact calculation and bonding

In the minimization step, the system is brought to a low energy state through molecular mechanics simulations (MM). In the following step, an MD-Simulation

**Fig. 10.28** Cross-linked
periodic unit cell of EPN.
Reaction sites are colored
*blue*, all other atoms are
colored *gray*. Note that the
network extends beyond the
original cell (*upper left*) to
the surrounding virtual cells

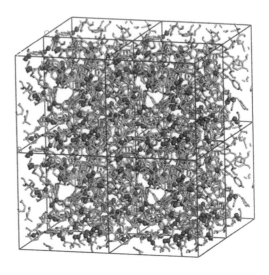

is performed to allow the reactive sites to close in on each other, guided by the artificial charges of opposite sign. In the third step, the distance of opposite reactive sites is calculated and, if in "close contact", a bond is formed. The close contact calculation is a function available in Materials Script of the Materials Studio software, where the distance $r$ between two atoms is considered "close" if smaller than a factor $C_{close}$ of the sum of the respective van der Waals radii $R_{i,vdw}$:

$$r \leq C_{close} \cdot \left( \left| R_{1,vdW} - R_{2,vdW} \right| \right) \tag{10.12}$$

For our investigations, we have found that a factor of $C_{close} = 1.1$ leads to sufficiently cross-linked packages while not producing too many artifacts (see below). After bonding, the involved reactive sites are taken from the list and the artificial charges are removed. Restarting the loop, this artificial bonding procedure is followed by the geometry optimization step necessary to avoid excessive potential energy of the system. The loop is rerun until termination conditions are met. Thereafter, all remaining reactive centers are saturated with hydrogen, which is thought reasonable since in reality highly reactive sites would find reaction partners e.g. from impurities to saturate. Finally, all artificial charges are removed and the partial charges of the whole structure recalculated. Figure 10.28 shows a detailed model of cross-linked EPN constructed in the way described above, emphasizing the reaction sites and true 3D-networking.

### 10.3.2.2 Validation of Atomistic Packing Models

The above described cross-linking procedure does not simulate the chemical process of cross-linking, but presents a way of constructing 3D-networked models that best represent the bulk structure of the epoxy. Considering the artificial

**Fig. 10.29** 2D-Sketch of
fully cross-linked models
representing nanograins (**a**)
and a network extending to
infinity (**b**)

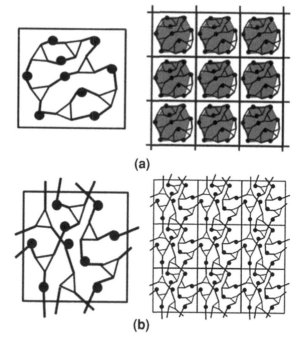

charges, the choice of reaction partners by close contact calculation and the fact that cross-linking occurs at a high temperature (for this system) and that it is an exothermic reaction, it becomes clear that this "quick and dirty" approach cannot claim to reproduce the cross-linking process itself, nor any properties that are involved (curing temperature, conversion rates, cure shrink etc.). Moreover, care has to be taken to check the resulting systems for artifacts such as ring spearing (a bond formed between atoms on opposite sides of an aromatic ring with the bond going through the center of the aromatic ring), "Olympic ring" catenation (where epoxy and BPA form a ring that physically interlocks with a similar structure) or other energetically unlikely structures. Some of the artifacts may be disputable, however, it is beyond the scope of this work to discuss all artifacts that may happen during the automated cross-linking routine in detail.

Another important detail of packing models that represent bulk epoxy resins is the three-dimensional nature of the network that consists, ideally, of one single molecule. Consider a simulation cell packed with a stoichiometric mixture of epoxy and hardener. It is in principle possible to react all cross-linking sites to get 100% conversion within this original simulation cell. However, if 3D-PBC are applied and the original cell is surrounded by virtual neighbors, the resulting structure would represent a nano-granular polymer rather than a 3D-network, because no connections between neighboring cells are formed.

**Fig. 10.30** NpT-MD
validation run of a packing
model at 298 K. The *gray
line* shows the typical frame-
to-frame density fluctuation
in a pressure-controlled
molecular dynamics run. The
*black line* depicts the running
average

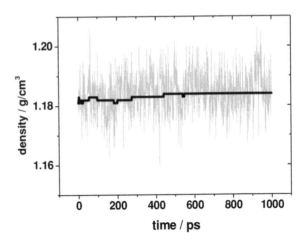

Figure 10.29 provides a sketch in two dimensions to emphasize the importance of 3D-connectivity for representing bulk behavior. In this work, we tested for connectivity using the *select_fragment* option of the materials visualizer, displaying the original cell surrounded by its 26 nearest virtual neighbors. 3D connectivity was considered sufficient, if not more than three (heavily entangled) fragments existed, not counting the odd unreacted BPA-monomer.

Several independent packing models were constructed this way: a batch of smaller size (monomer ratio of 4:6:18) resulting in packages of approximately 1,380 atoms and 24 Å edge length and a batch of larger size (8:12:36) resulting in packages of 2770 atoms and 30 Å, the exact number of atoms depending on the number of dangling ends. These packing models were subjected to an extensive equilibration procedure to make sure that the construction history of the packages is erased. The procedure involves an annealing step (molecular dynamics at constant number of atoms $N$, volume $V$ and temperature T: NVT-MD) at elevated temperature (600 K) of 1 ns and cooling down (NVT-MD) at 10 K/100 ps to 300 K. Velocities were then reassigned (Boltzmann distribution) according to a temperature of 298 K. After a short NVT-MD run of 10 ps, the ensemble was switched to NpT (fixed pressure $p = 0.1$ MPa) to let the volume fluctuate for 1 ns. After this equilibration procedure one more analysis run was performed (NpT-MD at T = 298 K and 1 ns duration) to assess the stability of the density as the last validation criterion of the package models. In Fig. 10.30 a representative validation run shows the stability of the packing model at a density of 1.184 g/cm$^3$.

Table 10.7 contains construction properties of the packing models along with the average densities of the validation run. Note that not all following investigations were performed on each of the packing models. The package named novosf2 of the smaller batch had to be discarded after construction due to a minimization failure caused by cross-linking artifacts.

**Table 10.7** Construction properties of bulk EPN models (target density 1.18 g/cm$^3$) and resulting equilibrated density and energy state

| Name of package | #atoms – | Edge length Å | Conversion (%) | Density g/cm$^3$ | Energy J/mol |
|---|---|---|---|---|---|
| Novosf1 | 1384 | 24.0 | 88.9 | 1.163 | 0.94 |
| Novosf3 | 1384 | 24.0 | 88.9 | 1.170 | 1.06 |
| Novosf4 | 1384 | 24.0 | 88.9 | 1.169 | 1.08 |
| Novosf5 | 1386 | 24.0 | 88.1 | 1.167 | 1.2 |
| Novolf1 | 2770 | 30.2 | 87.5 | 1.180 | 1.09 |
| Novolf2 | 2770 | 30.2 | 87.5 | 1.175 | 1.05 |
| Novolf3 | 2770 | 30.2 | 87.5 | 1.184 | 1.13 |
| Novolf4 | 2768 | 30.2 | 88.9 | 1.178 | 1.09 |
| Novolf5 | 2766 | 30.2 | 90.3 | 1.185 | 0.85 |

### 10.3.2.3 Bulk Properties of EPN

The equilibrated packing models of EPN have been subjected to two types of bulk-property analyses: (1) the estimation of elastic constants and (2) the estimation of the coefficient of thermal expansion (CTE).

1. The calculation of the elastic properties was performed by the discover module of the Materials Studio software of Accelrys. In short, each undeformed configuration is submitted to a minimization. Following this initial stage, three tensile and three pure shear deformations are applied and the system is re-minimized following each deformation. The internal stress tensor is then obtained from the analytically calculated virial and used to obtain estimates of the six columns of the elastic stiffness coefficients matrix. Assuming isotropic material properties, as is reasonable for amorphous polymers, the Lame constants $\lambda$ and $\mu$ can be obtained and the elastic constants Young's modulus E, bulk modulus B, shear Modulus G and Poisson's ratio $\nu$ derived [32]. A detailed discussion is addressed in the literature [43]. The analysis was performed on three packages of each batch of EPN (small and large) by arbitrarily choosing three configurations from the validation run. The results are summarized in Tables 10.2 and 10.3 and will be discussed below.

2. To get estimates of the CTE, heating runs were performed. After the validation run, the packing models were subjected to a series of NpT-MD runs, elevating the temperature by 10 K every 100 ps, covering the range of 300 K–500 K. Before each temperature increase, the specific volume $V_s = 1/\rho$ is recorded. From a plot of the change in specific volume $\Delta V_s/V_{s,0}$ against temperature T shows the thermal expansion of the system which, at the glass transition temperature, shows a change in slope (CTE). Since these are extensive and CPU intensive calculations, they were performed on all four small systems but only on one large one to check for improvements. Figure 10.31 shows the averaged results with error bars indicating the volume fluctuation during an NpT-MD run.

**Fig. 10.31** Volume expansion in heating runs of bulk EPN packages. From the slope of the curve in the glassy region (*dotted line*), and the rubbery region (*dashed line*), the volumetric expansion coefficient can be extracted. The *solid red line* indicates experimental data as already shown in Fig. 10.6

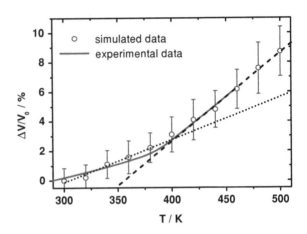

**Table 10.8** Calculated properties of EPN structures (small batch)

| Name of package | E/GPa | G/GPa | K/GPa | $v$ |
|---|---|---|---|---|
| Novosf1 | 3.929 | 1.488 | 3.640 | 0.320 |
| Novosf2 | – | – | – | – |
| Novosf3 | 4.654 | 1.763 | 4.302 | 0.320 |
| Novosf4 | 4.420 | 1.671 | 4.190 | 0.323 |
| Novosf5 | 1.587 | 0.555 | 3.786 | 0.430 |
| Average | 4.334 | 1.641 | 4.044 | 0.321 |
| ± std | ± 0.370 | ± 0.140 | ± 0.354 | ± 0.002 |

**Table 10.9** Calculated properties of EPN structures (large batch)

| Name of package | E/GPa | G/GPa | K/GPa | $v$ |
|---|---|---|---|---|
| Novolf1 | 3.839 | 1.404 | 4.819 | 0.367 |
| Novolf2 | 4.406 | 1.663 | 4.210 | 0.325 |
| Novolf3 | 4.341 | 1.604 | 4.921 | 0.353 |
| Novolf4 | 4.268 | 1.590 | 4.507 | 0.342 |
| Novolf5 | 4.142 | 1.548 | 4.281 | 0.339 |
| Average | 4.199 | 1.562 | 4.548 | 0.345 |
| ± std | ± 0.224 | ± 0.097 | ± 0.316 | ± 0.016 |

Tables 10.8 and 10.9. summarize the results of the numerically calculated elastic properties. Between all investigated models, the Young's Modulus varies only slightly with one exception in the small batch (novosf5), which has been excluded from the averaging procedure; it is suspected that insufficient 3D-connectivity leads to this obvious deviation. The averaged value of 4.26 GPa for the Young's modulus is in reasonable agreement with experimental data from dynamic mechanical analysis (DMA) at 1 Hz of 2 GPa (see Fig. 10.7). The elastic moduli are calculated from geometry-optimized (i.e. potential-energy minimized) static structures. These can be regarded as values valid in the absence of kinetic

energy, i.e. at T = 0 K. In light of that, the deviation of only a factor of 2 from the experiment can be seen as quite satisfying.

Similar reasoning holds for the calculation of the thermal expansion coefficient. In the glassy region, that is, at temperatures below about 400 K, no longer range relaxations or movements of side groups are expected in the real material and thermal expansion should be mainly due to the increased equilibrium distance of atom–atom interactions. However, in the rubbery region the polymer backbone and side groups become more mobile and therefore more contributions to the thermal expansion lead to an increased slope in the expansion-temperature plot. At or rather around the kink in the curve, the glass transition of the epoxy occurs.

In this way the data in Fig. 10.31 can be interpreted. In the low temperature region, the EPN packages show an average volume expansion coefficient of $\alpha_g = 280$ ppm/K (dotted line). Even at these very short simulation times (100 ps) the increased coefficient in the higher temperature region can be observed quite well. Here a value of $\alpha_r = 600$ ppm/K can be extracted from the simulations (dashed line). The simulated values of the CTE were derived by linearly fitting the five lowest (by temperature) and highest data, respectively. To compare simulated thermal expansion to experimental values, TMA measurements were carried out in expansion mode according to IPC-norm as shown in Fig. 10.6. In a cycle from room temperature to 470 K at 5 K/min, the third heating run was recorded and the relative length change $\Delta L/L_0$ converted into volume change assuming isotropic behavior:

$$\Delta V/V_0 = (1 + \Delta L/L_0)^3 - 1 \qquad (10.13)$$

The simulated value for the CTE in the glassy region agrees reasonably well to the experimental one of $\alpha_{g,exp} = 193$ ppm/K, deviating by a factor of 1.45. Experimental glass transition was obtained as $T_{g,exp} = 383$ K from the intersection of the fits, the glass transition temperature $T_g$ of the simulation can be derived yielding a value of $T_g = 403$ K. In the literature, values of 330 K for low conversion up to 395 K for highest (technical) conversion are reported, depending on the degree of cure [27]. Since the glass transition is also rate dependent (in this simulation: 20 K/100 ps), the calculated value may be considered as an upper boundary for the glass transition. However, for the highly cross-linked packing models, the deviation of 5% is very satisfying. The thermal expansion in the rubbery regime shows a remarkable agreement to the experimental value of $\alpha_{r,exp} = 570$ ppm/K. In Fig. 10.31, the experimental TMA-curve nearly matches the dashed line, which is the fit through the data of 420 K–500 K. While we suggest to treat the simulated data with care, keeping in mind the relatively high fluctuations, this result clearly shows the predictive capabilities of detailed molecular modeling.

### 10.3.2.4 Moisture Mobility

The equilibrated packing models have been investigated with respect to the mobility of water molecules. To that end, real samples were subjected to a typical temperature/humidity test condition of 60°C/60%r.h.. The samples were regularly

**Fig. 10.32** Mean squared displacement of water molecules in EPN, showing no regime of random walk

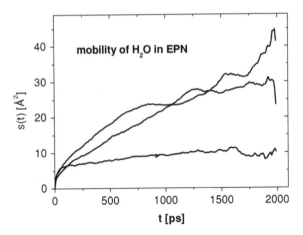

weighted and from the increase of mass, the saturation concentration of $C_{sat} = 0.9$ %wt. was determined (see Fig. 10.11). For the larger batch of packing models, this value amounts to 10 $H_2O$ molecules per packing model. The Materials Studio software provides a sorption tool that enables the insertion of a fixed number of penetrant molecules at energetically favorable sites of a package. Thus prepared models were subjected to molecular dynamics simulations (NpT ensemble) at 333 K, to investigate the mobility of the water molecules. Results of the mean square displacement $s(t)$ are plotted against simulation time in Fig. 10.32 for three models of EPN.

Obviously, no regime of random walk according to Eq. 10.6 and as discussed in Sect. 10.3.1.2 is reached within simulation time: On indication to reach this conclusion is the low value of $s(t) < 36$ Å$^2$, which computes to a displacement of merely $d = s(t)^{-0.5} = 6$ Å, which is barely above the estimated jump-distance of $a = 5$ Å (see Fig. 10.21). Vice versa, if the estimated jump distance is taken as basis along with the experimentally determined diffusion coefficient of $D = 1.8 \times 10^{-8}$ cm$^2$/s and a simulation time of 6 ns the number of diffusive jumps according to the Einstein relation Eq. 10.8 computes roughly to $n = 2$ for the investigated models. Since this is clearly insufficient to apply the random walk condition it must be concluded that for this system and at this temperature, the computational effort to observe diffusion of water exceeds our capabilities for the system of EPN.

### 10.3.3 Outlook: Interface Modeling

A layered 2D-periodic model of silicon-dioxide (SiO$_2$) was created based on a 3D-model from the Materials Studio database. Open bonds at the surfaces of this layer were saturated with hydroxyl-groups (-OH). Density stability of the thus constructed layer was checked in a short NpT-MD run.

**Table 10.10** Construction properties of the SiO₂–EPN interface models including equilibrated interfacial energies

| Package name | #atoms EPN/total | Edge length/Å X,Y/Z | Conversion/% | $E_{int}$/mJ/m² |
|---|---|---|---|---|
| Sio2_novo2 | 1667/2381 | 21.39/57.96 | 90.5 | 174 |
| Sio2_novo3 | 1681/2395 | 21.39/57.96 | 85.7 | 211 |
| Sio2_novo4 | 1681/2395 | 21.39/57.96 | 85.7 | 183 |

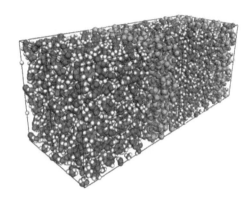

**Fig. 10.33** Interface model of EPN/SiO₂ in representation as van der Waals spheres. Coloration: carbon: *gray*; hydrogen: *white*; oxygen: *red*; silicon: *orange*. Displayed are the original cell and a virtual replica in *z*-direction, emphasizing the multi-layer aspect under 3D-PBC

A vacuum layer was added and filled with the EPN monomer mixture as described in the section for bulk models on page 16, using the amorphous cell packing task. The cross-linking algorithm and equilibration procedure was performed as described above.

In this way, three EPN–SiO₂ interface models were created containing a layer of SiO₂ with 20 Å thickness and a cross-linked EPN layer of 40 Å. It has to be noted that all four interface models share the same SiO₂ layer, while the EPN layers are independent of each other. Further, it has to be noted that the monomer mixture packed by the Amorphous Cell module is not strictly stoichiometric, however, the deviation in favor of reaching the target density of the EPN layer was deemed negligible.

Construction properties of the interface packages can be found in Table 10.10 and a typical representation of an interface model can be seen in Fig. 10.33. After optimizing the packages interfacial energies $E_{int}$ were calculated according to Yarovsky [25]:

$$\gamma = \frac{\Delta E}{2A} = [E_{SiO2} + E_{EPN} - E_{tot}]/2A \qquad (10.14)$$

where A = 21.4 Å² is the area of the interface, $E_{tot}$ the single point potential energy of the whole system and $E_{SiO2}$ and $E_{EPN}$ those of the silicon-oxide and EPN layers, respectively. Note that with 3D-PBC applied, the models are infinitely layered in the direction perpendicular to the interface (Fig. 10.33), and hence two interface areas have to be taken into account by Eq. 10.14.

**Fig. 10.34** The original
interface model of EPN
(*gray*) and SiO$_2$ (*red–orange*) has been copied and
divided by material.
Subsequently, the empty
space has been filled with
H$_2$O (*red–white*) molecules
to perform interfacial energy
analysis (see [44])

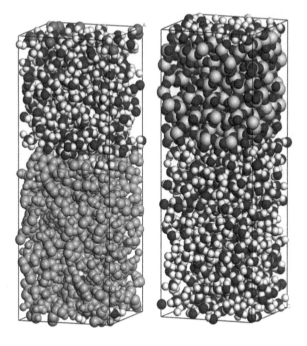

The calculated values for the interfacial energy densities are summarized in
Table 10.10. Positive values indicate thermodynamic stability, as is expected for
typical organic-adhesive/metal-oxide interfaces [25]. In their investigations of an
Aluminum/Epoxy interface, Yarovsky and Evans [8] obtained values of $\gamma_{int} = 0.1$
to 0.7 J/m$^2$ by simulation, indicating that the values in Table 10.10 meet expectations. The values of interfacial energies can be compared with work of adhesion
results from contact angle measurements.

In a contact angle measurement, a drop of a liquid is placed on the smooth
surface of a material. Depending on the balance of surface tensions $\gamma_L$, $\gamma_S$ and
interaction $\gamma_{LS}$ between liquid $L$ and solid $S$, the liquid drop forms an angle with
the solid surface. Their relation is given by the Young equation:

$$\gamma_L \cos\theta = \gamma_S - \gamma_{SL} \tag{10.15}$$

Knowing the properties of three different liquids, the contact angle data can
be evaluated to obtain the work of adhesion $w_{ad}$. For a detailed discussion see
[45–49].

To enhance the possibilities of comparison, the molecular models can be utilized to calculate the interfacial work of adhesion not only between the two solid
surfaces SiO$_2$ and EPN, but also of that to the different liquids. This has been done
in a recent publication of our group [44]. To that end, the convenient possibility in
molecular modeling to copy models and delete selected atoms or groups of atoms
is exploited. By way of Eq. 10.4, an energy sum is calculated which comprises all
bonded and non-bonded contributions according to the force field used. Thus in

**Fig. 10.35** Work of adhesion $w_{ad}$ calculated by molecular modeling for the liquids water ($H_2O$), methylene-iodide (MI) and glycerol (Glyc) on surfaces of EPN (*solid gray*) and $SiO_2$ (*hatched*) and between these solid surfaces (*leftmost bar*)

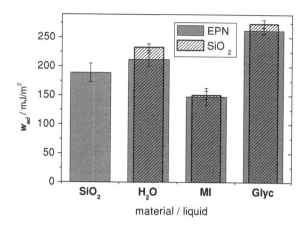

Eq. 10.14, all intra-material contributions (bonded or non-bonded) cancel out, while only interfacial contributions remain (which are of course non-bonded in nature). Note that in a 3D-periodic model without vacuum layer, again two interfaces need to be regarded and layer thicknesses need to be larger than potential cutoffs in order to avoid interactions across two layers. As an example, Fig. 10.34 shows this approach for $H_2O$ molecules inserted into the divided packages of EPN and $SiO_2$. From these packages, interfacial energies were calculated according to Eq. 10.14, with results for three different liquids presented in Fig. 10.35 (water ($H_2O$); methylene-iodide (MI); glycerol (Glyc)). The results obtained agree reasonably well with contact angle measurements, which are in the range of 80–150 mJ/m$^2$.

## 10.4  Conclusions

We have investigated epoxy resins with systematically varied composition to study structure–property correlations with respect to mechanical, structural and moisture transport properties. The work has been carried out on two different polymer systems as they are employed in electronic packaging, using simple experimental methods from classical materials testing as well as molecular dynamics simulation for an extensive parametric study.

We have found that trends could be predicted well, allowing identification of key features of the structure which influence the respective quantity under investigation. This underpins the introspective qualities of molecular dynamics simulation to create understanding of the underlying principles and mechanisms that can be described, studied and varied. In this vein, we have found:

- A simple branched approach enables quick assessment of properties and reveals trends as function of structural variables and loading conditions.
- More elaborate cross-linked models show results that agree with experimental results reasonably good.

- Free volume, polarity and matrix mobility are key descriptors of epoxy resin structure for diffusion and possibly also other properties.
- Transport of moisture can be analyzed but needs special attention regarding applicability of the respective equations. Very dense structures require extensive computational time.
- Interfacial work of adhesion seems a promising method to be assessed by detailed modeling paving the way to understand interface properties such as interface diffusion or delamination.

Quantitatively, the results show in part only reasonable agreement. This is, however, due to many factors that are difficult to assess and are based on the nature of a model and also its computational representation: Exact depiction of the amorphous material, its structure and chemistry, exact description of the interactions, representativeness of the volume element: all are based on best knowledge and numerical algorithms. This will, however, be only an approximation to the real situation. Further research, refined analytical methods and more computational power will surely significantly advance this rather new field in material science for electronic packaging, making molecular dynamics simulation a valuable tool to understand and describe properties and failure of materials and interfaces based on their structure within a multiscale framework towards thermo-mechanical lifetime prediction.

**Acknowledgments** We would like to thank our coauthors and colleagues that contributed to the publications [50–52] on which this paper is based. We also acknowledge financial support of parts of these works as indicated in the respective publication.

# References

1. Fan XJ, Zhang GQ, Driel WD van, Zhou J (2007) In: Zhang GQ, van Driel WD, Fan XJ (eds) Mechanics of Microelectronics, Springer, Netherlands, pp 281–375
2. Fan XJ (2008) Mechanics of moisture for polymers: fundamental concepts and model study. In Proceedings of Eurosime conference
3. Zhou J, Lahoti SP, Sitlani MP, Kallolimath SC, Putta R (2005) Investigation of non-uniform moisture distribution on determination of hygroscopic swelling coefficient and finite element modeling for a flip chip package. In: Proceedings of Eurosime Conference, pp 112–119
4. Zhou J (2006) Analytical and numerical bound analysis of hygroscopic swelling characterization. In: Proceedings of 56th ECTC, pp 734–739
5. Shirangi MH, Fan XJ, Michel B (2008) Mechanism of moisture diffusion, hygroscopic swelling and adhesion degradation in epoxy molding compounds. In: Proceedings of 41st international symposium on microelectronics (IMAPS), pp 917–923
6. Stellrecht E, Han B, Pecht MP (2004) Characterisation of hygroscopic swelling behaviour of mold compounds and plastic packages. IEEE Trans CPT 27:499–505
7. Vanderhart DL, Davis GT, Shen MA (1999) Partitioning of water between voids and the polymer matrix in a molymercompound by proton NMR: the role of larger voids in the phenomena of popcorning and delamination. J Microcircuits Electron Packag 22:424–439
8. Karad SK, Jones FR (2005) Mechanisms of moisture absorption by cyanate ester modified epoxy resin matrices: the clustering of water molecules. Polymer 46:2732–2738

9. Apicella A, Migliaresi C, Nicodemo L, Nicolais L, Iaccarino L, Roccotelli S (1982) Water sorption and mechanical properties of a glass-reinforced polyester resin. Composites 13(4):406–410

10. Soles CL, Chang FT, Bolan BA, Hristov HA et al (1998) Contributions of the nanovoid structure to the moisture absorption properties of epoxy resins. J Polym Sci Part B Polymer Phys 36:3035–3048

11. Luo S, Leisen J, Wong CP (2002) Study on the mobility of water and polymer chain in epoxy and 1st influence on adhesion. J Appl Polymer Sci 85:1–8

12. Duda JL, Zielinski JM (1996) Diffusion in Polymers. In: Neori P (ed) Free volume theory, Marcel Dekker, New York

13. Van Landigham MR, Eduljee RF, Gillespie JW (1999) Moisture diffusion in epoxy systems. J Appl Polymer Sci 71:787–798

14. Nogueira P, Ramirez C, Torres A, Abad MJ et al (2001) Effect of water sorption on the structure and mechanical properties of an epoxy resin system. J Appl Polymer Sci 80:71–80

15. Wu C, Xu W (2006) Atomistic molecular modelling of cross-linked epoxy resin. Polymer 47:6004–6009

16. Yarovsky I, Evans E (2002) Computer simulation of structure and properties of cross-linked polymers: application to epoxy resins. Polymer 43:963–969

17. Wong CKY, Fan HB, Yuen MMF (2008) Interfacial adhesion study for sam induced covalent bonded copper–EMC interface by molecular dynamics simulation. Compon Packag Technol IEEE Trans 31:297–308

18. Yang S, Gao F, Qu J (2010) A study of highly cross-linked epoxy molding compound and its interface with copper substrate by molecular dynamic simulations. In: 60th proceedings of electronic components and technology conference (ECTC), pp 128–134

19. Wu C, Xu W (2007) Atomistic molecular simulations of structure and dynamics of cross-linked epoxy resin. Polymer 48:5802–5812

20. Wu C, Xu W (2007) Atomistic simulation study of absorbed water influence on structure and properties of cross-linked epoxy resin. Polymer 48:5440–5448

21. Hofmann D, Fritz L, Ulbrich J, Schepers C, Böhning M (2000) Detailed-atomistic molecular modeling of small molecule diffusion and solution processes in polymeric membrane materials. Macromol Theory Simul 9:293–327

22. Hofmann D, Ulbrich J, Fritsch D, Paul D (1996) Molecular modelling simulation of gas transport in amorphous polyimide and poly(amide imide) membrane materials. Polymer 37:4773–4785

23. Fan HB, Chan EKL, Wong CKY, Yuen MMF (2006) Investigation of moisture diffusion in electronic packages by molecular dynamics simulation. J Adhesion Sci Technol 20:1937–1947

24. Lin YC, Chen X (2005) Investigation of moisture diffusion in epoxy system: experiments and molecular dynamics simulations. Chem Phys Lett 412:322–326

25. Yarovsky I (1997) Atomistic simulation of interfaces in materials: theory and applications. Aust J Phys 50:407–424

26. Deng M, Tan VBC, Tay TE (2004) Atomistic modeling: interfacial diffusion and adhesion of polycarbonate and silanes. Polymer 45:6399–6407

27. Jansen KMB et al (2004) Constitutive modeling of moulding compounds [electronic packaging applications]. In: Proceedings of 54th electronic components and technology conference pp 890–894

28. Jansen KMB, Wang L, van t Hof C, Ernst LJ, Bressers HJL, Zhang G Q (2004) Cure, temperature and time dependent constitutive modeling of moulding compounds. In: Proceedings of the 5th international conference on thermal and mechanical simulation and experiments in microelectronics and microsystems, EuroSimE 2004, pp 581–585

29. Shen CH, Springer GS (1976) Moisture absorption and desorption of composite materials. J Compos Mater 10:2–20

30. Kruse J et al (2003) Free volume in polyimides: positron annihilation experiments and molecular modeling. Macromolecules 38:9638–9643

31. Hinchliffe A (2003) Molecular modelling for beginners. Wiley, Chichester

32. Accelrys Software inc.(2009) Materials studio release notes, Release 5.0, San Diego

33. Sun H (2011) COMPASS: An ab initio force-field optimized for condensed-phase applications-overview with details on alkane and benzene compounds. J Phys Chem B 102(38):7338–7364

34. Einstein A (1908) Elementary theory of the brownian motion. Zeitschrift für Elektrochemie und Angewandte Physikalische Chemie 14:235–239

35. Goldreich P, Mahajan S, Phinney S (1999) Order of magnitude physics: understanding the world with dimensional analysis, educated guesswork, and white lies, University of Cambridge

36. Chen Z, Gu Q, Zou H, Zhao T, Wang H (2007) Molecular dynamics of water diffusion inside an amorhous polyacrylate latex film. J Polymer Sci 45:884–891

37. Bondi A (1968) Physical properties of molecular crystals liquids and glasses. Wiley, New York

38. Heuchel M, Böhning M, Hölck O, Siegert MR, Hofmann D (2006) Atomistic packing models for experimentally investigated swelling states induced by CO2 in glassy polysulfone and poly(ether sulfone). J Polymer Sci Part B Polymer Phys 44:1874–1897

39. Polymer User Guide, amorphous cell section,version 4.0.0 (1996) Molecular simulations, San Diego

40. Hofmann D, Entrialgo-Castano M, Lerbret A, Heuchel M, Yampolskii Y (2003) Molecular modeling investigation of free volume distributions in stiff chain polymers with conventional and ultrahigh free volume: comparison between molecular modeling and positron lifetime studies. Macromolecules 36:8528–8538

41. Hofmann D, Heuchel M, Yampolskii Y, Khotimskii V, Shantarovich V (2002) Free volume distributions in ultrahigh and lower free volume polymers: comparison between molecular modeling and positron lifetime studies. Macromolecules 35:2129–2140

42. Hölck O (2008) Gas sorption and swelling in glassy polymers combining experiment, phenomenological models and detailed atomistic molecular modeling. Federal Institute for Materials Research and Testing, Berlin

43. Theodorou DN, Suter UW (1986) Atomistic modeling of mechanical properties of polymeric glasses. Macromolecules 19:139–154

44. Hölck O, Bauer J, Wittler O, Michel B, Wunderle B (2011) Comparative characterization of chip to epoxy interfaces by molecular modeling and contact angle determination. In: 12th international conference on thermal, mechanical and multiphysics simulation and experiments in microelectronics and microsystems, EuroSimE 2011, Linz, Austria

45. Pocius AV (2002) Adhesion and adhesives technology. Hanser Gardner Publications, München, Germany

46. Van Oss CJ, Good RJ, Chaudhury MK (1998) Additive and nonadditive surface tension components and the interpretation of contact angles. Langmuir 4:884–891

47. Van Oss CJ, Chaudhury MK, Good RJ (1988) Interfacial Lifshitz-van der waals and polar interactions in macroscopic systems. Chem Rev 88:927–941

48. Comyn J (1992) Ontact angles and adhesive bonding. Int J Adhesion Adhesives 12:145–149

49. Fowkes FM (1962) Determination of interfacial tensions, contact angles and dispersion forces in surfaces by assuming additivity of iintermolecular iinteractions in surfaces. J Phys Chem 66:382

50. Dermitzaki E, Wunderle B, Bauer J, Walter H, Michel B, Reichl H (2008). In: Proceedings of 9th EuroSimE conference

51. Wunderle B et al (2009) Molecular dynamics simulation and mechanical characterisation for the establishment of structure-property correlations for epoxy resins in microelectronics packaging applications. In: 10th international conference on thermal, mechanical and multi-physics simulation and experiments in microelectronics and microsystems, EuroSimE, pp 1–11

52. Hölck O, Dermitzaki E, Wunderle B, Bauer J, Michel B (2010) Molecular modelling of microelectronic packaging materials—basic thermo-mechanical property estimation of a 3D-cross-linked epoxy/SiO2 interface, thermal, mechanical & multi-physics simulation, and experiments. In: 11th international conference on microelectronics and microsystems, EuroSimE, pp 1–10

# Part V
# Multiscale Methods and Perspectives

## Introduction

One ultimate vision of many modelers is the demonstration of complete scaling between the atomistic and the device application. Although the stretch or jump between scales is too aggressive to be practical today, one may begin the scaling issue by investigation of bridging techniques. As mentioned previously, Chap. 4 (by Shimokawa) discussed that one way to bridge the scaling issue is to apply quasi-continuum methods which embed molecular techniques. As a transition from the Shimokawa's work, this section begins in Chap. 11 (Fan et al.) with discussion of how molecular interfacial models may be directly applied to cohesive-zone continuum models, including an example of an interfacial copper-epoxy molecular model from which the force-displacement curve may be simulated. Chapter 12 (Chan et al.) discusses the strategy of larger-scale molecular models to simulate contact angle wetting and further applying the molecular contact-angle information to build finite element models describing aspects of surface roughness on the contact angle. Chapter 13 (Tesarski et al.) discusses model scaling using the typical functional-group-based mesoscale methods employed in the industry looking for density trends in the mesoscale model to abstract glass transition predictions. The methods are sufficiently new and need more investigation. However, Chap. 14 (Iwamoto) deals with direct scaling from the atomistic using a large-scale coarse-grained mesoscale bead targeted at epoxy-copper interfaces of interest in molding compounds. The significance of this work is that the scaling is taken to a larger level than usually used in molecular-mesoscale methods in order to simulate the modulus of both the epoxy and the epoxy-copper oxide interface, benchmarked against the measured modulus of the epoxy. Normal mesoscale methods use chemical functional groups as the basic parameterized bead units, however this work uses entire repeat units which were parameterized using molecular models, demonstrating that larger-scale jumping can be achieved.

# Part V Chapter List

**Chapter 11:** "Investigation of Interfacial Delamination in Electronic Packages"
H. Fan and M. M. F. Yuen

**Chapter 12:** "A Multiscale Approach to Investigate Wettability of Surfaces with Designed Coating"
E. K. L. Chan, H. Fan and M. M. F. Yuen

**Chapter 13:** "Glass Transition Analysis of Crosslinked Polymers: Numerical and Mesoscale Approach"
Sebastian J. Tesarski and Artur Wymyslowski

**Chapter 14:** "Investigation of Coarse-Grained Mesoscale Molecular Models for Mechanical Properties Simulation, as Parameterized Through Molecular Modeling"
Nancy Iwamoto

# Chapter 11
# Investigation of Interfacial Delamination in Electronic Packages

**H. Fan and M. M. F. Yuen**

This paper is based upon "A multiscale approach for investigation of interfacial delamination in electronic packages" by Haibo Fan and Matthew M. F. Yuen which appeared in Proceedings of Eurosime © 2007 IEEE.

**Abstract** In this study, a multiscale approach was proposed to study delamination in a bi-material structure, which bridge molecular dynamics method and finite element method using cohesive zone model. Cohesive zone model parameters were derived from an interfacial molecular dynamics model under mechanical loading and were assigned to the cohesive zone element representing the interfacial behavior. Based on the multi-scale model, the material behavior at nanoscale was passed onto the continuum model under tensile loading condition.

## 11.1 Introduction

Electronic package design is now moving not only toward high speed and multi-functional application, but also high density packaging with high performance requirements. Interfacial delamination, due to the presence of dissimilar material systems, is one of the primary concerns in an integrated circuit (IC) package design. The mismatch in the coefficient of thermal expansion between two different layers in IC packages can generate high interfacial stresses when

H. Fan (✉) · M. M. F. Yuen
Department of Mechanical Engineering, Hong Kong University of Science and Technology, Clear Water Bay, Kowloon, Hong Kong (SAR), China
e-mail: HB.FAN@philips.com

N. Iwamoto et al. (eds.), *Molecular Modeling and Multiscaling Issues for Electronic Material Applications*, DOI: 10.1007/978-1-4614-1728-6_11, © Springer Science+Business Media, LLC 2012

subjected to thermal loading during fabrication, which can result in delamination and destroy the functionality of the system.

An interface is a more complicated region that separates two non-miscible materials containing chemical bonding, and roughness. So far, it is still rather difficult to describe interface in one model considering all these effects. Traditional numerical method, like finite element method, is not suitable for modeling the chemical effect on the interfacial behavior. Molecular dynamics (MD) simulation is a well-established tool for modeling the material performance at an atomistic level including modulus, adhesion, thermal conductivity, solubility, diffusion and reactivity [1–8]. However, MD model is only suitable for modeling systems consisting of up to several thousands of atoms. The layout of a real structure always consists of the interfaces at different length scales from several nanometers to several millimeters or larger. It is impossible to build the full model using MD technique due to long calculation times and costly calculations. Multiscale modeling methods is still a challenge because of the different length scales and time scales involved in the models. Several methodologies on how to couple nano-scale models and continuum models for studying material performance of composites have been established, including hand-shaking method [9], coarse-grained molecular dynamics (CGMD) method [10] and virtual internal bond (VIB) method [11–15]. Hand-shaking method introduces displacement boundary conditions in interfacial region between the MD and finite element analysis (FEA) regions, where FEA mesh in the coupling region was scaled down to match the lattice of atomic cell. However, it is not easy to implement computational technique in the coupling region due to the higher distortion under large deformation, especially for the amorphous structure.

CGMD method seamlessly couples the MD regions to the continuum region through a statistical coarse graining procedure. However, the application of the multiscale method still suffers from mismatch of time scale occurring at the different length scales. VIB approach proposed by Gao and Klein [11] reproduces the behavior of a hyper-elastic solid, in which there are microstructures consisting of internal cohesive bonds based on the extension of the Cauchy–Born concept. VIB model can model crack nucleation and propagation without any presumed crack path in complex materials. However, VIB incorporates cohesive bonds into a constitutive law for the homogenized material particles. It is suitable for the bulk materials rather than description of atomic interaction along a prescribed interface. Moreover, VIB is based on the simplified atomic potentials without considering bond torsion, bending and electrostatic force, which is obviously not adequate to model the complicated reality of the material at atomistic scale across the material interface.

Molecular modeling endeavors to simulate the basic origins of material performance in a wide variety of topics including mechanical, chemical and electrical properties. With proper atomic description relative to the measurement (energy potential, structure and environmental conditions), the reliable information could be extracted from MD simulation and adequately represent the material response being measured, such as, mechanical modulus for specific low $k$ dielectric spin-on materials [3] and for epoxy resin materials [6]. Therefore, it is possible to propose a hierarchical multiscale method incorporating the information obtained by MD simulations into the continuum model to investigate the constitutive response of bulk composite which contains nano materials. The methodology does provide an indication that information of interfacial failure at nano scale could be transferred to traditional continuum models by cohesive element.

In this study, a multiscale approach was proposed to study delamination in a bi-material structure, which bridge molecular dynamics method and finite element method using cohesive zone model. Cohesive zone model parameters were derived from an interfacial MD model under mechanical loading and were assigned to the cohesive zone element representing the interfacial behavior. Based on the multiscale model, the material behavior at nanoscale was passed onto the continuum model under tensile loading condition.

## 11.2 Computational Methodology

The interfacial failure is an adhesion problem which is governed by the interfacial bonding, in particular the molecular bonding across the interface. Except for atoms belonging to bulk materials attached to the interface, interface is covered by some other atoms like oxygen atoms or molecules like water molecules, as well as chemical bonds formed among these interface atoms. Moreover, rough surfaces at the atomic scale represent the nature of the interface, where large gaps exist. Obviously, these dominant factors at atomic scale govern interfacial adhesion rather than bulk material properties of two bonded materials. Kendall [16] also found that the adhesion between surfaces is dominated by a number of factors such as van der Waals force, chemical bonding and surface roughness. Obviously, without considering all these issues at the interface, continuum model is not enough to simulate interfacial delamination. In spite of long calculation times and costly calculations in MD simulation, MD models can easily and explicitly provide the interfacial behavior of a local area under different mechanical loading conditions considering chemical treatment at the interface, such as bond broking, defect generation and delamination propagation. Therefore, MD simulation can provide traction force under the applied displacement during interface separation, which is the basis of the cohesion model for interfacial delamination. It is indicated that a multiscale investigation from atomic simulation to continuum simulation could be established for complete understanding of interfacial delmaination.

An atomic-based continuum model will be proposed to investigate interfacial delamination in this study, as illustrated in Fig. 11.1. An interfacial MD model will be built to find the constitutive relation of the interface under external mechanical loads. A continuum FEA model is built with cohesive zone elements laid on the interface and the constitutive relations from interfacial MD model is input to cohesive elements to simulate interfacial delamination under the mechanical

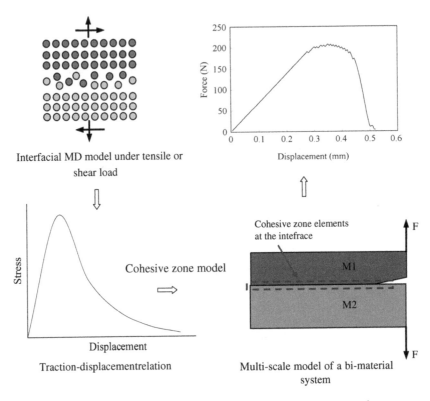

Interfacial MD model under tensile or shear load

Traction-displacementrelation

Cohesive zone model

Multi-scale model of a bi-material system

**Fig. 11.1** An illustration of the proposed model linking nanoscale and macroscale

loading. The corresponding failure force varying with the applied displacement will be extracted from the model, which can be used to guide experiment for interface material design.

## 11.2.1 Interfacial MD Model

From the engineer's standpoint, interface always bears interfacial stresses coming from the bulk bodies bonded to the interface. These stresses are the macroscopic collective behavior of the atomistic bond network and govern the crack nucleation and propagation of the interface. Therefore, it is rather important to derive the constitutive relation of the interface (stress–displacement relation) from MD simulations.

Normally interfacial MD mode is built with a rectangular simulation box in the $x$ and $y$ directions, periodically located in the plane perpendicular to the interface, as shown in Fig. 11.2. A large vacuum space is positioned at the top of the model

**Fig. 11.2** An illustration of interfacial MD model of bi-material system

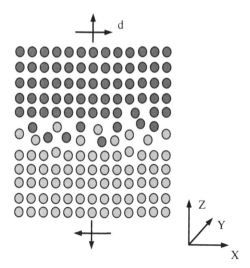

in order to avoid interaction across the mirror image in the $z$ direction in the calculations. Energy minimization is first performed to find the equilibrated structure of the bi-material system. Then all the atoms except for those two layers of atoms near the interface are held rigid in all simulations. A tensile or shear displacement is applied on the model in single simulation step and the displacement is maintained by the time interval for the relaxation of the system before the same next displacement is applied. The above MD procedure is repeated until the interface is completely separated. The atomic configurations and energies of the system for each simulation step are monitored and recorded during the simulations. The simulations are conducted by using Discover module of the Materials Studio software (Accelrys Inc.) COMPASS force field enables accurate prediction of material properties for a broad range of materials under different conditions. The COMPASS force field can accurately be applied on the systems of polymers, metals and their interfaces.

Normally, atoms in MD simulations are modeled as point masses interacting through potentials, which are usually characterized experimentally. The potential energy of the system provides the forces on each atom, which can be used to determine the acceleration, velocity and positions of each atom. In the classical molecular dynamics method, the equations of motion for atoms are described by Newton's equations as follows:

$$F_i = m_i \frac{d^2 r_i}{dt^2} \quad F_i = -\nabla_i \Phi, \tag{11.1}$$

where $F_i, m_i$ and $r_i$ are respectively the force vector, mass and position vector of molecule $i$. $\Phi$ is the potential energy function of the system.

**Fig. 11.3** Constitutive
relation for the interface

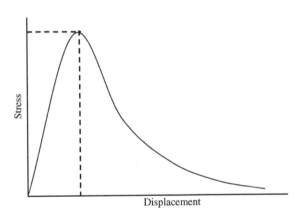

When the bi-material system is subjected to external displacement, force will be transferred to the interface by the interaction among atoms whose position and velocity are governed by above potential energy. The corresponding interfacial stresses can be calculated as follows:

$$\sigma_{\alpha,\beta} = \frac{1}{A} \sum_i \left( \sum_j \frac{\partial \Phi}{\partial (r_{ij})} (r_{ij})_\alpha \cdot (r_{ij})_\beta \right), \tag{11.2}$$

where $\sigma_{\alpha\beta}$ is interfacial component, $\alpha$ and $\beta$ correspond to the $x$, $y$ and $z$ directions, $A$ is the interfacial area.

Stress–displacement relation can be derived from MD simulations, as shown in Fig. 11.3. The relation shows the nonlinear behavior of the interface, increasing stress first and then decreasing stress with the increasing displacement. This is a constitutive cohesive relation of interface, describing the relation of interfacial traction force and opening displacement during delamination propagation. As presented by Shet and Chandra [17], the cohesive curve starts from point, A, where interface starts to separate, reaches point, B, where cohesive crack tip is located, and finally comes to point, C, where interface is completely separated.

Cohesive zone model (CZM), is widely used to simulate fracture process in different kinds of composites under different loading conditions [18]. The key of the CZM is the traction–displacement constitutive relation representing interfacial fracture behavior. However, it is still rather difficult to experimentally determine these parameters due to complex interfacial adhesion governed by molecular bonds and roughness. In this proposed method, these key parameters are derived by MD simulations, which avoid some experimental issues.

CZM has been widely used to study fracture process because of avoiding singularity at the crack tip and easy implementation in traditional FEA models. In a cohesive zone model, energy is allowed to flow into the fracture process zone for surface separation. Normally cohesive relation is described by cohesive

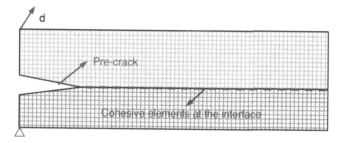

**Fig. 11.4**  An illustration of multi-scale model of bi-material system

parameters, namely cohesive strength, $\sigma_{max}$, separation distance, $\delta$, and cohesive energy, $\phi$, derived by the area under the traction–displacement curve. These cohesive parameters could be obtained from the above constitutive relation derived by MD simulations as shown in Fig. 11.4, which normally constitute cohesive zone models with linear [19], bilinear [20], trapezoidal [21] and exponential shape [22], respectively. The shape of the cohesive zone model has some effects on the analysis of interfacial delamination [23]. Bilinear and exponential cohesive models are selected and implemented in commercial codes ANSYS and ABAQUS.

### 11.2.2 Cohesive Zone Model

A multiscale model of a bi-material system was built by using the ANSYS code, as shown in Fig. 11.4, to study the interfacial delamination of the bi-material system under mechanical loading. In this model, the both materials are modeled as continuum with homogeneous and elastic properties. Solid element is used to model material 1 and material 2. Cohesive zone elements are laid on the interfaces except for a part of the interface where a pre-crack is made, as shown in Fig. 11.4. Cohesive element is used at the interface to describe the behavior with the selected cohesive zone model. Figure 11.5 shows the schematic of planar cohesive element. The initial thickness of the underfomed element is set to zero and the interfacial separation is defined as displacement jump, $\delta$, the difference of the displacement of the adjacent interfacial nodes for deformed element. The relation of nodal force and interfacial separation is governed by the cohesive relation derived by MD simulations. Under external mechanical loads, the system undergoes elastic deformation and total energy is bared by elastic energy and cohesive energy dissipated within the cohesive elements. The cohesive energy goes within the crack tip region to separate the interface. When new free surfaces were created, the traction force and the stiffness of the cohesive zone elements on these free surfaces go to zero, but the displacement across them is still continuous. That is why cohesive zone model can be implemented in FEA model for interface separation without loss of continuity.

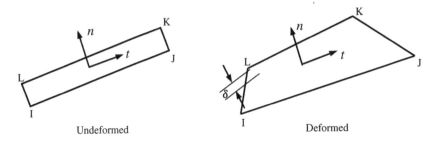

**Fig. 11.5** Schematic of undeformed and deformed cohesive element

**Fig. 11.6** The plot of
fracture force and
displacement applied on the
model

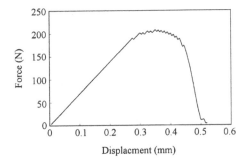

With stipulated external displacement applied to the model, the corresponding failure force under the applied displacement is extracted from the multi-scale model during interface separation, as shown typically in Fig. 11.6. The maximum fracture force is used to estimate the fracture strength of the interface which characterizes the interfacial material property.

## 11.3 Case Study

In this study, MD simulations are performed to evaluate adhesion between an epoxy molding compound (EMC) and a copper substrate. MD model includes a fragment of EMC and copper atoms. The fully cured epoxy network is composed of diglycidyl ether of bisphenol-A (DGEBA) epoxy and methylene diamine dianilene (MDA) curing agent and the model is same as that presented by Fan et al. [8]. The model did not include solid components such as filler and pigments, which would require large scale models that are beyond the current MD simulation capability. Based on the same method, the fully cured epoxy network was layered with a cuprous oxide surface cleaved from a crystal structure, corresponding to the (001) plane. The cured epoxy chains were initially placed on the substrate. A large vacuum spacer was positioned at the top of the epoxy chains in order to avoid

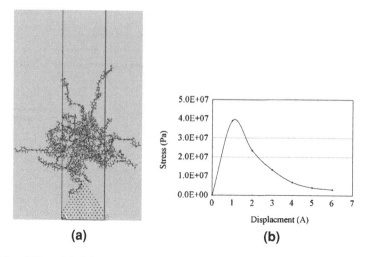

**Fig. 11.7 a** MD model of the EMC–Cu system. **b** Stress–displacement relation for the EMC–Cu system

interaction across the mirror image in the $z$ direction in the calculations. The MD models were built with a rectangular simulation box $3.53 \times 3.53$ nm$^2$ in the $x$ and $y$ directions, periodic in the plane perpendicular to the EMC–Cu interface. All the copper atoms were held rigid, while all the EMC chains were allowed to move freely in all simulations. Energy minimization was performed to find the equilibrated structure of the bi-material system using the ensemble of the constant number of particles, constant-volume and constant temperature (NVT) at 25°C. Fig. 11.7a shows the morphological configuration with the minimum potential energy for the MD model.

Based on the above procedure proposed, the constitutive cohesive relation of the EMC/Cu interface is derived from MD simulations, as shown in Fig. 11.7b. The curve showed that the nonlinear behavior of the interface, increasing stress first and then decreasing stress with the increasing displacement. The curve provides the cohesive parameters for cohesive elements.

Tapered double cantilever beam (TDCB) test is carried out to evaluate the tensile adhesion between EMC and copper substrate. In order to study the interfacial delamination of EMC and Cu substrate under mechanical loading, it is necessary to perform finite element analysis to extract some useful information. A more realistic multi-scale model of TDCB test is shown in Fig. 11.8a. The mesh was refined at the interface between the EMC and copper to capture the steep stress gradients expected. A pre-crack with a length of 39 mm is made at the interface. Cohesive elements are used at the EMC/Cu interface and the initial thickness of the cohesive elements is set to zero. Both EMC and Cu materials are assumed to be linear elastic, homogeneous and isotropic. The constitutive relations derived from the interfacial MD model as shown in Fig 11.7b is assigned to the cohesive zone elements as the description of the atomic interaction between the

**Fig. 11.8  a** The diagram of
the TDCB assembled with
EMC and copper leadframe.
**b** Force–displacement curve
for the EMC–Cu system

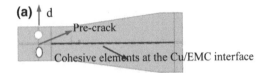

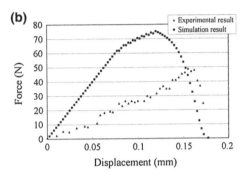

EMC and copper substrate. The displacements of nodes at the surface of the hole
in the bottom block were constrained and tensile displacement was applied on the
surface of the top hole.

The tensile force was calculated for the multiscale model under the tensile displacement and plotted against the displacement, as shown in Fig. 11.8b. The tensile
force increased with the increased displacement and reached the maximum value
where delamination initiated, and then decreased to zero for the remainder of the
displacement. The higher maximum tensile force means the higher adhesion between
EMC and Cu substrate. Predicted result from TDCB test simulations showed that the
adhesion force between EMC and SAM treated Cu substrate was higher than that for
the control sample. Experimental result is also shown in Fig 11.8b. It can be seen that
the predicted result from simulations were a little bit higher that the experimental
result. The difference between the simulation and experimental value can be
attributed to more complicated crosslink density of EMC. Moreover, voids or
impurity inside the real samples can also degrade the interfacial properties.

## 11.4  Summary and Discussion

A simple and effective multiscale approach was proposed to study delamination
in a bi-material structure, which bridge molecular dynamics method and finite
element method using cohesive zone model. With proper formulation of the
MD model and appropriate use of boundary conditions, potential functions and
simulation procedure, MD simulation can provide good understanding of delamination at fundamental level, and the parameters of the cohesive zone model for
delamination propagation.

In contrast to other multi-scale methods, the method presented in this study has significant advantages. It avoids the complicated numerical equations to solve the overlapping domain in the method involving coupling of continuum models with molecular models. We also demonstrated a methodology to investigate delamination initiation at the epoxy molding compound (EMC)/Cu interface [15], in which the interfacial material properties were derived from atomic force microscopy (AFM) measurements using the Lennard–Jones potential. In that study, we considered only van der Waals force because the adhesion between the EMC and copper was dominated by non-chemical bonding interactions. However, that method is not suitable any more in the above case due to the complicated interfacial bonds between EMC and copper substrate. In this approach, the atomistic behavior is directly transferred from the nanoscale to the continuum scale by the constitutive relation derived MD simulations, therefore it avoids the suffering from mismatch of time scale occurring at the different length scales. It can also predict the material behavior more accurately than VIB method considering simplified potential energy.

A bifurcation-based multi-scale decohesion model was developed by Shen and Chen [24] to investigate delamination between tungstem film and silicon substrate. They conducted MD simulations to obtain decohesion relation of single crystal W block under tensile loading and implemented the model into the material point method (MPM). However, MPM is the method for the size scaling down the continuum level to the atomic level, so size of the model is still within the nano scale. Moreover, they also argued that the proposed model should be verified by an integrated experimental, analytical and numerical investigation on the structures with sizes varying from nano scale to macro scale.

Namilae and Chandra [25] also developed a hierarchical multiscale method to study interfacial shear strength between CNT and its polymer matrix by the cohesive zone model parameters. This model had successfully employed to study the effect of interfacial strength on the elastic properties of the composites. However, they did not provide any experimental evidence for the model that the atomistic behavior of the interface from the MD model was successfully passed on to the continuum model.

Based on the method presented in this study, the atomistic information including deformation, void nucleation and interfacial debonding were extracted and represented by the constitutive relation. The constitutive relation of the interface of the epoxy resin polymer and Cu substrate was derived from MD simulations under tensile stain and is assigned to the TDCB model to calculate the tensile forces. The predicted results were found to be comparable with those from experimental measurement, which indicates that the proposed approach can be used to study delamination at the interface consisting of nanoscale materials.

**Acknowledgments** This study was financially supported by the Grant Research Founding 621907.

# References

1. Tanaka G, Goettler LA (2002) Predicting the bonding energy for nylon 6,6/clay nanocomposites by molecular modelling. Polymer 43:541–553
2. Gou J, Minaie B, Wang B, Liang ZY, Zhang C (2004) Computational and experimental study of interfacial bonding of single-walled nanotube reinforced composites. Comput Mater Sci 31(3–4):225–236
3. Iwamoto N, Moro L, Bedwell B, Apen P (2002) Understanding modulus trends in ultra low k dielectric materials through the use of molecular modeling. Proceedings of the 52nd electronic components and technology conference May 28–31, San Diego, CA, pp 1318–1322
4. Fan HB, Chan EKL, Wong CKY, Yuen MMF (2006a) Investigation of moisture diffusion in electronic packaging by molecular dynamic simulation. J Adhes Sci Technol 20:1937–1947
5. Fan HB, Chan EKL, Wong CKY, Yuen MMF (2007) Molecular dynamic simulation of thermal cycling test in electronic packaging. ASME J Electron Packag 129:35–40
6. Fan HB, Yuen MMF (2007) Material properties of the cross-linked epoxy resin compound predicted by molecular dynamics simulation. Polymer 48:2174–2178
7. Wong CKY, Fan HB, Yuen MMF (2008) Investigation of adhesion properties of Cu-EMC interface by molecular dynamics simulation. IEEE Trans Compon Packag Technol 31: 297–308
8. Fan HB, Zhang K, Yuen MMF (2009) The interfacial thermal conductance between a vertical single-wall carbon nanotubes and a silicon substrate. J Appl Phys 106:034307
9. Lidorikis E, Bachlechner ME, Kalia RK, Nakano A, Vashishta P, Voyiadjis J (2001) Coupling length scales for multiscale atomistics-continuum simulations: atomistically induced stress distributions in Si/Si3N4 nanopixels. Phys Rev Lett 87:086104
10. Rudd RE, Broughton JQ (2000) Concurrent coupling of length scales in solid state systems. Phys Status Solidi B 217:251–291
11. Gao H, Klein P (1998) Numerical simulation of crack growth in an isotropic solid with randomized internal cohesive bonds. J Mech Phys Solids 46:187–218
12. Klein P, Gao H (1998) Crack nucleation and growth as strain localization in a virtual-bond continuum. Eng Fract Mech 61:21–48
13. Ji B, Gao H (2004) A study of fracture mechanisms in biological nano-composites via the virtual interbal bond model. Mater Sci Eng A 366:96–103
14. Gao H, Ji B (2003) Modeling fracture in nanomaterials via a virtual internal bond method. Eng Fract Mech 70:1777–1791
15. Fan HB, Wong CKY, Yuen MMF (2006b) A multi-scale method to investigate delamination in electronic packaging. J Adhes Sci Technol 20:1061–1078
16. Kendall K (2001) Molecular adhesion and its applications: the sticky universe. Kluwer Academic/Plenum Publishers, New York
17. Shet S, Chamdra N (2002) Analysis of energy balance when using cohesive zone model to simulate fracture process. J Eng Mater Technol 124:440–450
18. Xu XP, Needleman A (1994) Numerical simulation of fast crack growth in brittle solids. J Mech Phys Solids 42:1397–1434
19. Camacho GT, Ortiz M (1996) Computational modeling of impact damage in brittle materials. Int J Solids Struct 33:2899–2938
20. Geubelle PH, Baylor J (1998) The impact-induced delamination of laminated composites: a 2D simulation. Compos Part B 29B:589–602
21. Tvergaard V, Hutchinson JW (1992) The relation between crack growth resistance and fracture process parameters in elastic-plastic solids. J Mech Phys Solids 40:1377–1397
22. Needleman A (1990) An analysis of decohesion along an imperfect interface. Int J Fract 42:21–40
23. Alfano G (2006) On the influence of the shape of the interface law on the application of cohesive-zone models. Compos Sci Technol 66:723–730

24. Shen L, Chen Z (2004) An investigation of the effect of interfacial atomic potential on the stress transition in thin films. Model Simul Mater Sci Eng 12:347–369
25. Namilae S, Chamdra N (2005) Multiscale model to study the effect of interfaces in carbon nanotube-based composites. J Eng Mater Technol 127:222–232

# Chapter 12
# A Multiscale Approach to Investigate Wettability of Surfaces with Designed Coating

E. K. L. Chan, H. Fan and M. M. F. Yuen

This paper is based upon "Multiscale approach optimization on surface wettability change", by E. K. L. Chan, H. Fan, and M. M. F. Yuen which appeared in the Proceedings of Eurosime 2010 © Year, IEEE.

**Abstract** Surface wettability is known to be governed not only by chemical structure but also by the surface geometrical structure. A multiscale approach on rough surface wettability was presented to study the combined effects of surface interactions. Different chemical structures and configurations were first input into the molecular model to get equilibrated structures. Contact angle was then estimated and input into continuum model with a roughness factor included.

## 12.1 Introduction

Surface wettability is one of the paramount properties of a solid surface that controls its interaction with liquid media. For a liquid on a flat surface, the contact angle is considered to be the combined result of surface tension at the tripartite solid/liquid/gas interface which can be described Young's equation [1]. Moreover, surface wettability is not only governed by chemical structure but also by the surface geometrical structure. In real applications, solid surfaces are usually not perfectly flat but are somewhat rough, so the effect of surface roughness has to be considered for surface wettability. Wenzel's equation and Cassie's equation are

E. K. L. Chan (✉) · H. Fan · M. M. F. Yuen
Department of Mechanical Engineering, Hong Kong University of Science and Technology, Clear Water Bay, Kowloon, Hong Kong (SAR), China
e-mail: edwardc@ust.hk

N. Iwamoto et al. (eds.), *Molecular Modeling and Multiscaling Issues for Electronic Material Applications*, DOI: 10.1007/978-1-4614-1728-6_12, © Springer Science+Business Media, LLC 2012

two main theories that describe the relationship between the apparent contact angle and surface roughness on solid surfaces [2–5].

To study the surface wettability change induced by surface chemical structure, photoresponsive material was employed on the surface. Photoresponsive materials include inorganic nanomaterials to small organic molecules and photoactive polymers. For organic complexes, there are several kinds of derivative compounds based on their photoactive groups, such as azobenzenes, spiropyrans, and cinnamates. Under photo-irradiation, the chemical configuration of these groups change between two states, with which the molecular polarity and surface free energy change accordingly, leading to a transition of surface wettability.

Azobenzene materials can be photo-switched to trans and cis isomers under light actuation. The different forms for isomers correspond to different dipole moments and surface wettability. Ichimura et al. [6] reported light-driven motion of liquids on a flat substrate surface modified with photochromic azobenzene units prepared by the chemisorption self-assembly technique. By the same technique, Feng et al. [7] fabricated an azobenzene polymer film through Langmuir–Blodgett (LB) technique, on which the change of contact angle is about 10°. Although the wettability of azobenzene has attracted great attention, the change of water contact angle of azobenzene so far reported is limited, almost no more than 10°. Meanwhile, comparing with the chemisorption and LB technique, electrostatic layer-by-layer self-assembly has been considered as a simple, versatile, and effective technique for fabrication of ultrathin organic films by alternately dipping substrates into dilute solutions of cationic and anionic polyelectrolytes [8]. They reported a hydrophobic electrostatic self-assembly azobenzene monolayer coated on a roughened silicon substrate surface resulting in a large reversible change of wettability (the maximum change of water contact angle about 66°) after UV and visible irradiation.

The introduction of roughness changes surface wettability significantly. The governing equations, however, cannot solely be described by Wenzel's or Cassie's equation easily. Recently, researchers [9–11] have used molecular dynamics (MD) calculations to study the influence of surface roughness on liquid droplet contact angle. Different chemical structures or configurations can be input into the molecular model to get equilibrium structures. However, these molecular models cannot reflect the real situation where roughness usually is in micrometer scale and large computational power is needed for surfaces with certain roughness.

On the other hand, finite element model is better posed to simulate the roughness effect on the contact angle by specifying the surface energy between liquid and substrate in ideally flat scenario [12–15]. Nevertheless, finite element model cannot simulate the behavior of photo-responsive materials at nanometer scale. Therefore, multiscale approach which incorporated the information obtained by MD simulations into the continuum model to investigate wettability of surface is becoming attractive.

In the following text, a multiscale approach was developed to study wettability of different surfaces coated with nanomaterials. MD simulations were conducted to obtain contact angles of nanomaterial coated surfaces, which were assigned to the continuum model consisting of water and rough surface. The roughness effect on the wettability was predicted for optimization of material and surface structure to control the hydrophobicity/hydrophilicity at liquid/solid interface.

## 12.2 Multiscale Simulation Methodology

Multiscale simulation approach was started by building the azobenzene material through commercial software Material Studio 4.0 (Accelrys Inc. USA). Two substrates, glass and silicon, and two types of azobenzene materials were selected. Trans and cis configuration of the azobenzene material was placed on top of the substrate respectively. Energy minimization was conducted to obtain the lowest energy state. Shape of water molecules was gathered after simulation and the contact angle is calculated from this equilibrium state. The contact angle information was then input into the computational fluid dynamics (CFD) FLUENT software, ANSYS Inc. Roughness was introduced on the glass and silicon surface in the form of rectangular pillars with different size and pitch configurations. The shape of water molecules was again gathered after the finite element simulation and the contact angle was estimated from this equilibrium state.

### 12.2.1 Molecular Dynamics Modeling

In order to predict the contact angle change by photoisomerization of azobenzene material on rough surface, Lennard–Jones equation was used for the solid–liquid interaction. Periodic boundary conditions were specified at the four side surfaces and a mirror boundary at the top surface. In this simulation approach, the simple point charge/extended potential and Lennard–Jones potential were used for water–water and water–substrate interactions, respectively. Different trans and cis configurations of the azobenzene material were input into the molecular model and their interfacial energy was calculated. Two azobenzene material candidates are involved in this study and their chemical structures are shown in Fig. 12.1.

In this simulation, the silicon surface was cleaved from a crystal silicon structure, corresponding to the (001) plane. As the non-bond cutoff distance in the force field setting is 9.5 Å, the depth of silicon surface used in the simulation was about 10 Å. A layer builder was used to build a sandwich and a large vacuum spacer was positioned at the top of the silicon surface in order to avoid interaction across the mirror image in the $z$ direction in the calculations. The MD models were built with a rectangular simulation box $5.02 \times 5.02$ nm$^2$ in the $x$ and $y$ directions, periodic in the plane perpendicular to the interface. The azobenzene material chains were initially placed on the silicon substrate. The azobenzene material

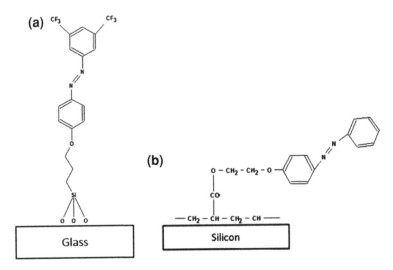

**Fig. 12.1** Chemical structures of **a** 4-hydroxy-3′,5′-bis(trifluoromethyl)azodibenzene, **b** Poly {2-[4-phenylazophenoxy]ethyl acrylate-co-acrylic acid}

chain is configured with trans or cis configurations. The Lennard–Jones potential is used for liquid–liquid interactions. In this simulation, the simple point charge/ extended potential [16] and Zhu–Philpott potential [17] were used for water–water and water–platinum interactions, respectively. Energy minimization was carried out to obtain the lowest energy state. All simulations were carried out at the temperature of 25°C, using the ensembles of the constant number of particles, constant-volume and constant temperature (NVT). All the simulations were performed with an interval of 1 femtosecond (fs) in each MD simulation step.

After the conducting conformation of azobenzene material on the substrate, a freely relaxed water drop consisting of 670 molecules was centered on top of the azobenzene material surfaces. The interactions among water molecules and between water molecules and azobenzene material surface consist of van der Waals force and Coulomb force. The parameters used in the simulations are from the polymer consistent force files (PCFF). A cutoff distance of 12.5 Å was used for these nonbonding interaction forces. All the simulations were conducted at 25°C using NVT for about 500 ps with an interval of 1 fs in each simulation step.

Figures 12.2 and 12.3 show the final configurations of water droplets on the azobenzene coated silicon and glass before and after UV treatment. Water droplet on the two substrates was spread finally with different spherical shapes of droplet cap. The contact angle for each scenario is estimated and listed in Table 12.1. Experimental results from sessile drop method were also listed in the table.

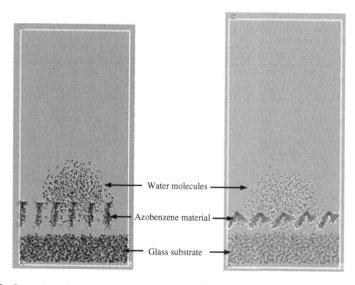

**Fig. 12.2** Contact angle of azobenzene material before UV (*left*) and after UV (*right*) on the glass substrate

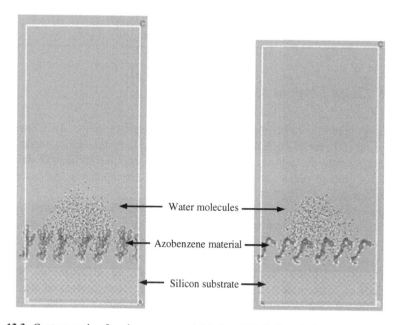

**Fig. 12.3** Contact angle of azobenzene material before UV (*left*) and after UV (*right*) on the silicon substrate

**Table 12.1** Contact angle results by molecular dynamics (MD) simulation and sessile drop contact angle measurement

|  | Glass (MD) | Glass (experiment) | Silicon (MD) | Silicon (experiment) |
|---|---|---|---|---|
| Before UV | 109° | 98° | 76° | 78° |
| After UV | 70° | 77° | 71° | 76° |

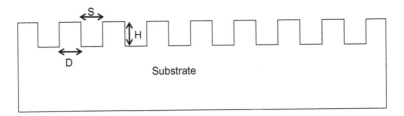

**Fig. 12.4** Pillar configurations for ANSYS model

**Table 12.2** Substrate configuration used in simulation

| Substrate material | Size (D) (μm) | Spacing (S) (μm) | Height (H) (μm) |
|---|---|---|---|
| Glass | 60 | 50, 100, 150, 200 | 30 |
| Silicon | 10 | 20, 40, 60 | 20 |

## 12.2.2 Finite Element Modeling

An axis symmetric volume-of-fluid (VOF) model is established and simulated by commercial software FLUENT 6.2. Roughed surface is designed for contact angle simulation by introducing rectangular pillars on the substrate surface, as shown in Fig. 12.4. Different pillar sizes, spacing and height are input into the model and their configuration is listed in Table 12.2.

Model configuration used was shown in Fig. 12.5. The axis-symmetric two-dimension (2D) solution domain was selected for saving computation time. Air and liquid phases were modeled as incompressible, Newtonian fluids with constant value of viscosity and surface tension. Water was used as liquid in the simulation with viscosity of 0.001 kg/ms, density of 998.7 kg/m$^3$, and surface tension with air of 0.073 N/m. In this model, it adopts an algorithm about segregated laminar flow, with a pressure inlet of 1 atm at the top boundary of atmosphere zone.

Because the free surface between air and water significantly changes shape and location during the course of VOF simulations, a uniform grid (with aspect ratio of unity) was used. Higher mesh density was used in the model with smaller pillar spacing to increase solution accuracy. Estimated contact angles from MD simulation (as listed in Table 12.1) were input into the model to simulate the effect of

**Fig. 12.5** Model setup of water contact angle

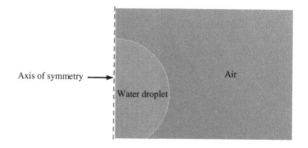

**Fig. 12.6** Contour of water volume fracture for cis-azobenzene material coated glass with 100 μm spacing

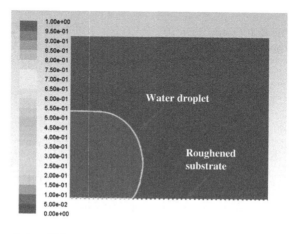

Contours of Volume fraction (water)   (Time=5.9997e-02)

surface roughness on the azobenzene material coating and to estimate the respective surface wettability for each individual trans and cis configuration. Contact angle was estimated by the steady-state solution obtained from the finite element model.

Figure 12.6 shows contour of water volume fracture, and water contact angle is obtained from the equilibrium state of water droplet for different pillar spacings were summarized in Figs. 12.7 and 12.8.

## 12.3  Summary and Discussion

From MD simulations, it is obviously seen that water contact angle for azobenzene material coated substrate is smaller at cis configuration (after UV). The photo-switched wettability is a reflection of the change in the dipole moment of the azobenzene unit upon trans to cis isomerization. While the hydrophobic CF3 group in  4-hydroxy-3′,5′-bis(trifluoromethyl)azodibenzene  results  in  less  interaction

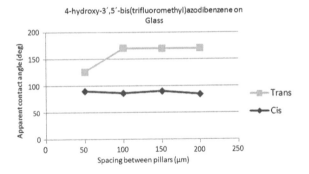

**Fig. 12.7** Contact angle transition of azobenzene material coated glass of different pillar spacings

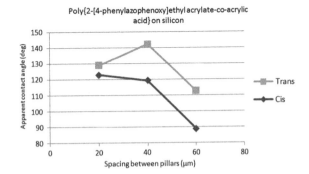

**Fig. 12.8** Contact angle transition of azobenzene material coated silicon of different pillar spacings

between water molecules and 4-hydroxy-3′,5′-bis(trifluoromethyl) azodibenzene, making a larger contact angle. When comparing results from sessile drop experiment, results are comparable with the same order and trend. Deviation may be from actual deposition density, alignment, and photoisomerization efficiency of azobenzene material on the substrate.

After inputting the results obtained from MD model into finite element model, contact angles of water droplets on different roughed substrates were obtained. For azobenzene material coated on glass, simulation results showed an increased contact angle when spacing is larger than 100 μm. On the other hand, a stable contact angle trend of cis-azobenzne was found with increasing pillar spacing. Increasing the spacing between pillars correlates the decreasing in surface roughness on the substrate. From the results shown in Fig. 12.6, water droplet sticks onto one pillar and cannot wet other pillars as the spacing increases. Theoretically, equilibrium of the contact line becomes possible only when the drop sits on pillar tops. The drop can sit on the air pocket but the contact line cannot. Thus, free displacement of the contact line becomes impossible. Water drop needs sufficient energy to jump from one pillar to another pillar. It explains why water cannot wet another pillar easily, which leads to insignificant change in contact angle with increase of pillar spacing. Also, cis-azobenzene material has a smaller contact angle compared with trans-azobenzene material with a change of around 80°. This means that the contact angle will have larger change before and after UV irradiation and this is confirmed by other research [7].

**Fig. 12.9**  Partial filling of water inside the grooves

Compared with finite element simulation results of azobenzene material coated on silicon with smaller pillar spacing as Jiang et al. [8], the same trend in our case was found coherently as shown in Fig. 12.8. From their results, superhydrophobicity on the monolayer was obtained by introducing <100 µm geometrical structure (square pillars) on silicon substrate surface and a largest reversible contact angle change was realized at 40 µm pillar spacing through UV and Vis-irradiation. By introducing sub-hundred micrometer pillars, an extraordinary trend of contact angle was found. This further confirms the importance of roughness value in terms of wettability change by azobenzene material. By comparing the glass and silicon case, a different scale of roughness leads to extraordinary wetting behavior.

It was found that in both scenarios the contact angle increased by introducing surface roughness. Air was trapped under the water droplet. Partial filling of water inside the grooves was found for different pillar spacings as shown in Fig. 12.9. However the filling factor was different in different scenarios. Hence, the contact angle could not be described by simple Wenzel or Cassie–Baxter empirical formulas. Nevertheless, the air pockets trapped under the water droplet is the predominant factor for increase in water contact angle to superhydrophobicity state; even no superhydrophobicity phenomenon was found on flat substrate in the MD models.

This multiscale approach provides an opportunity to study the combined effects of surface interaction from molecular scale to micron scale on the wettability of a rough surface. It enables the prediction of contact angle of liquid media on rough surfaces in the selection and optimization of material and surface structure to control the hydrophobicity/hydrophilicity at liquid/solid interface. By comparing the glass and silicon case, a different scale of roughness leads to extraordinary wetting behavior.

# References

1. Adamson AW, Gast AP (1997) Physical chemistry of surfaces. Wiley, New York
2. Wenzel RN (1936) Resistance of solid surfaces to wetting by water. Ind Eng Chem 28:988–994
3. Cassie ABD, Baxter S (1944) Wettability of porous surfaces. Trans Faraday Soc 40:546–551
4. Erbil HY, Demirel AL, Avcı Y, Mert O (2003) Transformation of a simple plastic into a superhydrophobic surface. Science 299:1377–1380

5. Han TJ, Lee DH, Ryu CY, Cho K (2004) Fabrication of superhydrophobic surface from a supramolecular organosilane with quadruple hydrogen bonding. J Am Chem Soc 126: 4796–4797
6. Ichimura K, Oh SK, Nakagawa M (2000) Light-driven motion of liquids on a photoresponsive surface. Science 288:1624–1626
7. Feng CL, Zhang YJ, Jin J, Song YL, Xie LY, Qu GR, Jiang L, Zhu DB (2001) Reversible wettability of photoresponsive fluorine-containing azobenzene polymer in langmuir–blodgett films. Langmuir 17:4593–4595
8. Jiang W, Wang G, He Y, Wang X, An Y, Song Y, Jiang L (2005) Photo-switched wettability on an electrostatic self-assembly azobenzene monolayer. Chem Commun 28:3550–3552
9. Maruyama S, Kimura T, Lu MC (2002) Molecular scale aspects of liquid contact on a solid surface. Therm Sci Eng 10(6):23–29
10. Yang C, Tartaglino U, Persson BNJ (2008) Nanodroplets on rough hydrophilic and hydrophobic surfaces. Eur Phys J E 25:139–152
11. Shi B, Dhir VK (2009) Molecular dynamics simulation of the contact angle of liquids on solid surfaces. J Chem Phys 130:034705
12. Unverdi SO, Tryggvason G (1992) A Front tracking method for viscous, incompressible, multi-fluid flows. J Comput Phys 100:25–37
13. Fukai J, Zhao Z, Poulikakos D, Megaridis CM, Miyatake O (1993) Modeling of the deformation of a liquid droplet impinging upon at surface. Phys Fluids A 5:2588–2599
14. Monaghan JJ (1994) Simulating free surfaces with SPH. J Comput Phys 110:399–406
15. Ranade VV (2002) Computational flow modeling for chemical reactor engineering. Academic Press, London
16. Berendsen HJC, Grigera JR, Straatsma TP (1987) The missing term in effective pair potential. J Phys Chem 91:6269–6271
17. Zhu SB, Philpott MR (1994) Interaction of water with metal surfaces. J Chem Phys 100:6961–6968

# Chapter 13
# Glass Transition Analysis of Cross-Linked Polymers: Numerical and Mesoscale Approach

Sebastian J. Tesarski and Artur Wymyslowski

**Abstract** Molecular modeling is one of the fastest developing tools in material science. There are a couple of reasons of such a state: on the one hand molecular modeling today seems to be much more user friendly, and on the other hand it is much more efficient in comparison to research based on traditional experiments, which are quite expensive and long lasting. Although the basic problem of numerical analysis is accuracy, in certain cases we can accept even high inaccuracy as long as the predicted tendency or trends is assessed properly. In recent years there has been a noticeable tendency and need for numerical material science using multiscale analysis especially in case of polymer materials. Thus, recently a number of researchers have concentrated on molecular mesoscale modeling of cross-link polymers. Cross-linked polymers seem to be very important in microelectronic and nanoelectronic packaging and assembly. One of the basic benefits of mesoscale analysis is the possibility of extending the time and length scale and reduce the usage of CPU power needed for analysis. In this paper we describe the preliminary research on cross-linked polymers and results of numerical modeling, which was done in Accelerys Material Studio and facilitated by its scripting capabilities through user defined subroutines. The developed subroutine allows one to differentiate statistically the process of polymer model creation and saves time needed for preparing the simulation. The main goal of the analysis was to estimate the glass transition temperature of the selected polymer through the density versus temperature dependence.

S. J. Tesarski (✉) · A. Wymyslowski
Faculty of Microsystem Electronics and Photonics,
Wroclaw University of Technology, Wroclaw, Poland
e-mail: sebastian.tesarski@pwr.wroc.pl

N. Iwamoto et al. (eds.), *Molecular Modeling and Multiscaling Issues
for Electronic Material Applications*, DOI: 10.1007/978-1-4614-1728-6_13,
© Springer Science+Business Media, LLC 2012

## 13.1 Introduction

### 13.1.1 Background

Since the 1960s electronic devices have been expected to work faster while becoming smaller, and at the same time they have been expected to be reliable. With the decreasing size of the devices new problems with their construction appear e.g. density of the current, heat dissipation, interference in communication lines. Electronic packaging encounters the similar problems. Usage of new materials, dedicated for such application is mandatory. Polymers, with their huge variety and range of values of physical properties, fit well in this trend. Polymer materials are used for encapsulation, underfills, for flip chip, moulding compound, electrically or thermally conductive adhesive, flexible electronics, materials for Printed circuits board (PCB), etc. Out of the vast number of polymers used in microelectronics, the main focus in packaging is directed towards the so-called cross-linked polymers.

As the experimental studies are long lasting and expensive, there is a tendency to use numerical tools for constructing, assessing and predicting the basic properties of selected cross-linked polymers. In the early 1950s of the twentieth century researches made some steps in order to model the world that surrounds them. The first work on molecular modeling was developed using plastic balls kept in big reservoirs and observation of their behavior for external interruption. In order to make more appreciative models the reservoirs had to be bigger and bigger, and such an approach consumed budgetary resources. Some researches started to implement equations describing the balls attached to each other in the first computers. At this time the machines were big and not sufficient, so only simple models, as in [1], could be simulated with great effort. Fortunately, computational power grew exponentially, and in a few years computers were capable of computing larger molecular systems. At first, every system had to be described by researcher with proper equations. With time computer aided design (CAD) tools were developed, so nowadays a user interface is more friendly and does not involve writing equations by hand. Users only have to choose which atom they want to connect to another atom and in this way the compound is built. Additionally, users give some properties to the atoms and from that information CAD tools are able to give answers to questions about properties of the molecular system.

### 13.1.2 CAD Tools

The market offers a vast number of CAD software packages in any engineering and/or research field. The authors have been particularly interested in software dedicated to modeling of materials in nano and microscale. Most of the programs

**Fig. 13.1** Numerical model
of Resin (**a**) and Hardener (**b**)

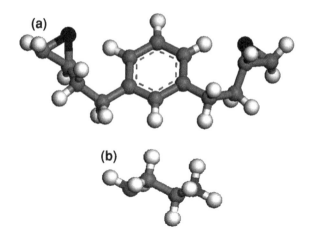

dedicated for materials science are developed for and by chemists. Authors of such programs assumed some level of chemical knowledge of the potential users, which is the key assumption. Materials Studio (Accelrys, USA) is used for molecular dynamics simulation in research.

Materials Studio offers 26-modules which are independent programs for specialized calculation. Modules can be divided into five groups: visualization, quantum tools, classical simulation tools, mesoscale simulation tools, analytical and crystallization tools and statistical tools. Materials Studio is a comprehensive materials modeling and simulation application designed for scientists in chemicals and materials R&D as well as in pharmaceuticals development. It provides a comprehensive set of scientific capabilities for modeling crystal structure and crystallization processes; for property prediction for molecules, polymers, catalysts, and other materials; and for the development of structure–activity relationships.[1]

## 13.2 Numerical Model Description: Chain Approach

According to the literature, one of the most popular cross-linked polymers among manufacturers of microelectronics industry is based on EPON 862 as a resin and triethylenetetramine (TETA) as a hardner. Based on data given in [2] an algorithm for creating the cross-linked network was developed. Authors described the algorithm in [3]. The developed models of hardener and resin are given in Fig. 13.1.

Materials Studio does not support a procedure of building cross-linked polymer from scratch, so the authors decided to write a dedicated subroutine in Perl scripting language included in Material Studio. The developed flowchart for that is illustrated

---

[1] www.accelrys.com

**Fig. 13.2** Flowchart of the developed cross-linking subroutine

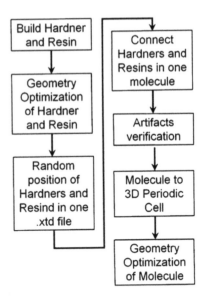

in Fig. 13.2. The authors used the idea given by Wu and Xu [4] and scripts developed by Fraunhofer IZM and Accelrys during the Nanointerface Project.[2]

According to the flowchart an appropriate model of hardener and resin is built and included into separated documents. The models are then optimized geometrically in order to assure proper space distribution of atoms in a molecule. After that a chosen amount of hardener and resin molecules are scattered randomly (different coordinates and different angles) in common space. Then hardener and resin molecules are connected in order to form one big molecule—the closer the active ends of hardener and resin molecules the bigger the probability of connection. After this step it is mandatory to check whether the compounds did not connect in a non-chemical way. In the following step a molecule is put into the 3D periodic cell, so that the boundary conditions are fulfilled. The last step is to apply geometry optimization for the whole constructed molecule.

In the first approach the authors decided to put an already cross-linked molecule into the 3D periodic cell rather than putting the individual resin and hardener molecules, due to the easier script implementation of the cross-linking procedure outside of the 3D Periodic Cell. Nevertheless, it is planned in the next step to apply cross-linking procedure after putting the separate molecules of hardeners and resins into the 3D periodic cell.

The authors are aware of possible problems due to the applied modeling procedure, especially as the proper periodic boundary conditions of cross-linking are not fulfilled. Nevertheless, it seems worthy to verify the developed procedure and to compare its results with the more appropriate cross-linking planned as a next step of the above research. The authors decided to compare the results for

---

[2] www.nanointerface.eu

**Fig. 13.3** Scheme of the
cross-linked molecule
including the so-called single
chain marked in the *red box*

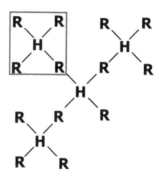

three molecules consisting of different numbers of basic chains as shown in
Fig. 13.3. For the current analysis, three molecules were constructed including 5,
10 and 15 chains, which are shown in Fig. 13.4a–c respectively. It was assumed
that the more chains included in the molecule the more accurate results could be
expected.

## 13.2.1 Atomistic Analysis

The algorithm of molecular dynamics simulation is partly based on the procedure
described precisely in [5, 6]. The first step is to equilibrate the structure in 298 K
for 300 ps using the ensemble of constant number of particles, volume and temperature (NVT). It is followed by the ensemble of constant number of particles,
pressure and temperature and NVT is used for calculating density in the range of
temperature from 513 K to 293 K in step of 10 K for 200 ps time for every
step. For the next iteration the previous velocities of atoms were used. Fig. 13.5
illustrates the applied algorithm of the following atomistic simulation steps.

   The aim of this calculation was to obtain volume versus. temperature depen-
dencies of the cross-linked polymer, which could be used to derive glass transition
temperature and coefficient of thermal expansion. Fig. 13.6 showed the change in
density throughout the temperature range for atomistic simulation where density
decreases with the increase of temperature for three kinds of chains. Fig. 13.7
shows the percentage volume change versus. temperature for the atomistic simulation, indicating that the volume increases with the increase in temperature.

## 13.2.2 Mesoscale Analysis

In the case of the mesoscale analysis approach, the first step was based on
changing the model from atomistic scale to mesoscale. The idea of converting
the atomic model into a meso model relies on replacing a few atoms with one
pseudo-atom called a bead. The lesser the atoms within one bead, the more

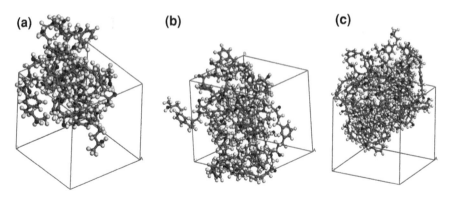

Fig. 13.4 The created molecules consisting of five chains (**a**), ten chains (**b**) and 15 chains (**c**)

**Fig. 13.5** Algorithm of the
applied atomistic simulation

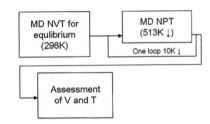

**Fig. 13.6** Density versus
temperature for atomistic
simulation

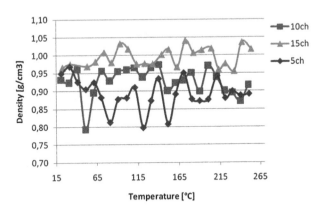

accurate it is, but also it increases the need for CPU Power. So the user must
balance between accuracy and available CPU power for simulation. The authors,
based on [7, 8], decided to change each resin molecule into three beads and
hardener molecules into two beads as shown in Figs. 13.8 and 13.9. The same
algorithm as in the previous section (Fig. 13.5) was used for assessing the density

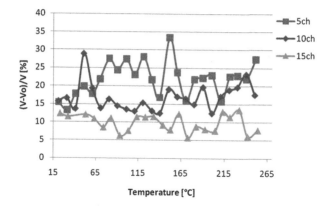

**Fig. 13.7** Volume versus temperature for atomistic simulation

**Fig. 13.8** The scheme of
converting Resin molecule
into three mesoscale beads

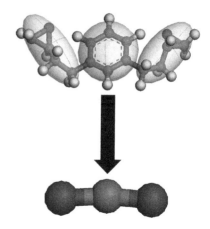

at certain temperatures. The only difference was the time length of the simulations.
For example, for the atomistic models in the previous section, 200 ps were used,
while for the mesoscale simulation 200 and 20,000 ps were used. Figure 13.10
shows
a final constructed mesoscale model for 5, 10 and 15 chain atomistic molecules.

   Similarly, as for the atomistic case, the aim of the calculation was to obtain
volume versus. temperature dependency of cross-linked polymer, which could be
used to derive glass transition temperature and coefficient of thermal expansion.
Figures 13.11 and 13.12 respectively, show the change of density versus. temperature for simulations at 200 and 20,000 ps. At both time lengths, the density
decrease with increase in temperature. It can be concluded from the achieved
that the results extending time length resulted in less scatter of the density values.
At the same time the trend of the line slope did not change significantly.

**Fig. 13.9** The scheme of
converting hardener molecule
into two mesoscale beads

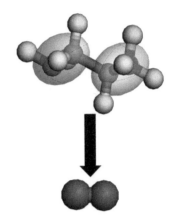

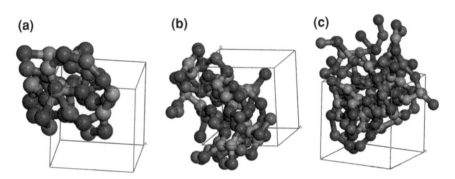

**Fig. 13.10** The created mesoscale molecues consisting of five chains (**a**), ten chains (**b**),
15 chains (**c**)

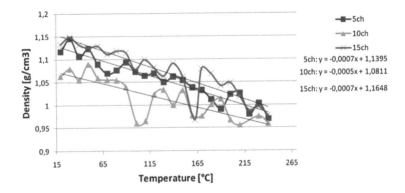

**Fig. 13.11** Density versus temperature for 200 ps mesoscale simulation

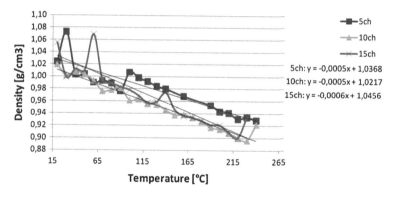

**Fig. 13.12**  Density versus temperature for 20,000 ps mesoscale simulation

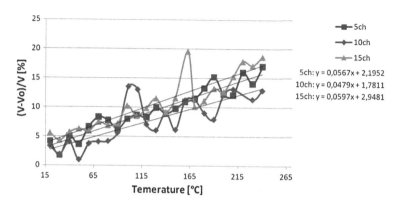

**Fig. 13.13**  Volume versus temperature for 200 ps mesoscale simulation

Figures 13.13 and 13.14 show the change of volume versus temperature for simulations at 200 and 20,000 ps. It was found that for both graphs the volume increases in the increase of the temperature. The trend of the line slope seems to be similar for both time lengths, while the resulting data for the 20,000 ps graph is less scattered than for the 200 ps graph.

In order to understand how conversion of the atomistic model into the meso-scale model influences the final results, the authors decided to design an additional numerical experiment, in which 14 models were prepared. Two kinds of hardeners (listed in Table 13.1) and seven different schemes of epoxy resins (listed in Table 13.2) were used in the simulations. All combinations gave 14 models in total. Every model was geometrically optimized. The idea was to divide to smaller beads with the assumption that every bead should contain the same amount of atoms and the atoms in the beads should be logically connected with each other. There was an additional assumption that one bead should not contain less than two atoms. Thus there are only seven possibilities for dividing the resin. Applying

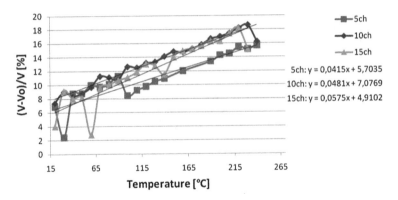

**Fig. 13.14** Volume versus temperature for 20,000 ps mesoscale simulation

| **Table 13.1** hardener divided into beads | Numbers of beads/bead types | Divided hardener |
|---|---|---|
| | 1/1 (1H) | |
| | 1/1 (2H) | |

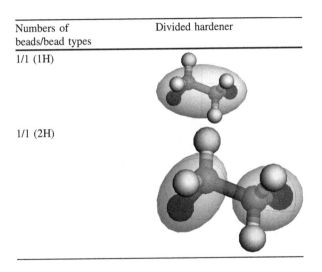

those rules gave us only two ways of dividing the hardener. In this study a five chain model was used.

The first step of modeling was geometry optimization, the second was calculation at 298 K NVT [(constant: number of atoms ($N$), volume ($V$) and temperature ($T$)] to relax the structure. The next step was to calculate the density and change in volume at certain temperature points. For this calculation NPT was used [constant: number of atoms ($N$), pressure ($P$) and temperature ($T$)]. The starting temperature was 513 K and the calculation was done for every 10 K, until 293 K. Figure 13.15 shows density as a function of temperature for all bead converting types. As it can be seen, the more the beads in the model the lower the value of density and the value of density is less scattered. In mesoscale simulation only 3R-1H and 3R-H2 models are near this value. So it can be assumed that the above conversion scheme is the best in this case.

**Table 13.2** Resin divided into beads

| Numbers of beads/bead types | Divided resin |
| --- | --- |
| 1/1 (1R) | |
| 2/2 (2R) | |
| 3/2 (3R) | |
| 5/3 (5R) | |
| 6/3 (6R) | |

(continued)

**Table 13.2** (continued)

| Numbers of beads/bead types | Divided resin |
|---|---|
| 11/4 (11R) | 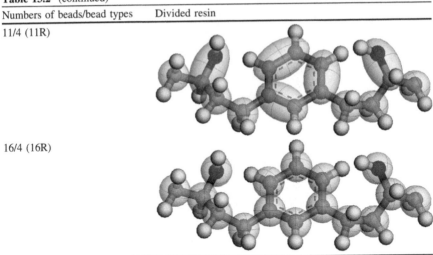 |
| 16/4 (16R) | |

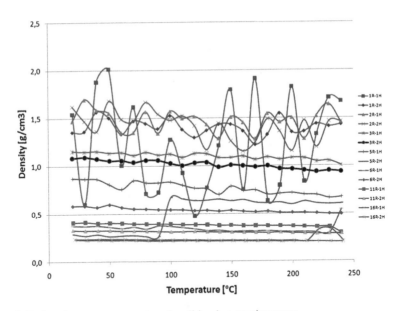

**Fig. 13.15** Density versus temperature for all bead converting types

## 13.2.3 Short Conclusion

It was noticed that as more chains were used the less scatter of the resulting data was found, though the final trend was not totally as expected. Comparing modeling results given on Figs. 13.6, 13.11, and 13.12, it would be difficult to define the

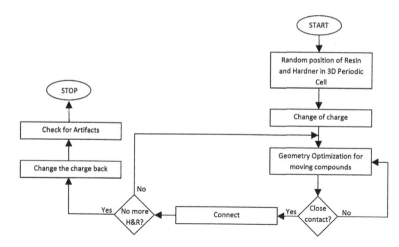

**Fig. 13.16** Algorithm of cross-linking in 3DPC

change in density. Analyzing simulation data it could be estimated that the glass transition temperature would be about 390 K although this may be due to proper interpretation, especially if the data can be well fitted by one or two lines with different slopes. The same applies to Figs. 13.7, 13.13, and 13.14; and it would be difficult to identify precisely the change of volume.

Mesoscale models, 3R-1H and 3R-2H, gave similar results as the experimental data and atomistic simulation; in these models the benzene ring is not divided into smaller beads. It can be concluded that the function groups should not be divided while using the beads instead of atoms. Additionally, there is a need for more profound analyses including more advanced cross-linking procedure with more accurate simulation, as presented in the next sections.

## 13.3 Numerical Model Description: Fully Cross-Linked Approach

In the previous approach the hardeners and the resins were cross-linked together before placing them into 3D Periodic Cells. They were cross-linked by the pattern found in Fig. 13.3. Due to boundary conditions in 3D Periodic Cells, models prepared in this way were seen as viscous fluids with separate molecules near each other in infinite rows and columns. The model was not seen as a segment of infinite cross-linked polymer. As can be seen from previous simulations, the chain approach does not give proper results. In order to change this state a new approach has been introduced.

The elaborated algorithm is given in Fig. 13.16. The first step was to place randomly placed resins and hardeners into the 3DPC. The next step was to increase

**Fig. 13.17** Example of possible artifacts

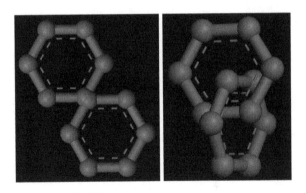

**Fig. 13.18** Algorithm of relaxing procedure using the module Forcite within Materials Studio

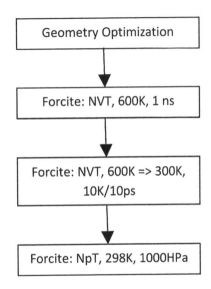

the charges of reactive ends in order to allow spontaneous flow of compounds. If two reactive ends are in close contact a connection was established. The process runs in a loop until all free resins and hardeners are gone. The next step was to restore the initial charges. The last step was to check for possible artifacts and connection that are physically impossible to happen in a real case. Figure 13.17 presents examples of possible artifacts.

After creating the model it was important to subject it to a relaxation procedure, in order to deprive the model from built-in stresses. An algorithm of such a procedure is presented in Fig. 13.18. The first model is geometry optimized; next it is annealed at 600 K for 1 ns. The next step is to decrease temperature to 300 K at a rate 10 K/10 ps. Finally the model was simulated at room temperature under one atmosphere in order to check stability of the density.

**Fig. 13.19** Algorithm of molecular simulation

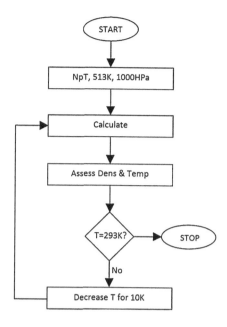

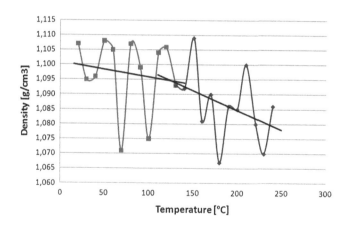

**Fig. 13.20** Density versus temperature fully cross-linked

## 13.3.1 Atomistic Analysis

After the relaxation procedure the model was ready for the final simulation. Figure 13.19 shows the algorithm for assessment of the density change as a function of temperature. Simulation starts at 513 K and decreases to 293 K with a decreasing temperature rate of 10 K. Density is derived at every temperature step and Fig. 13.20 shows the density change as a function of temperature. As it can be

**Fig. 13.21** The idea of
conversion

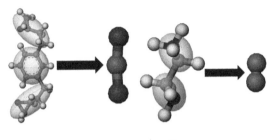

**Fig. 13.22** Model in
mesoscale

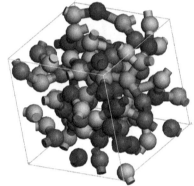

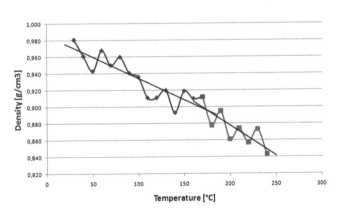

**Fig. 13.23** Density versus temperature for mesoscale model

seen points are quite scattered but the tendency could be observed. The possible
glass transition temperature is near its literature value.

## 13.3.2 Mesoscale Analysis

In order to create the mesoscale model, a fully cross-linked atomistic model was
used. Based on the research described previously, it was decided that the resin will
be divided into three beads and hardener into two beads. Figure 13.21 presents the

idea of conversion and Fig. 13.22 presents the model in mesoscale. The algorithm of simulation is the same as in the atomistic simulation (Fig. 13.19). Figure 13.23 shows density as a function of temperature for the fully cross-linked mesoscale model. It can be seen that data points are less scattered than those for the atomistic simulation but the tendency is weaker, which may due to the high glass transition temperature.

Finally it can be concluded that the new cross-linking algorithm, which mimics natural movement of compound during cross-linking, seems to give more accurate results. It is performed in 3DPC which means that for the simulation a segment from an infinite structure is taken.

## 13.4 Conclusions

This work presents the preliminary results of research directed towards numerical and experimental analysis of cross-linked polymers. The method presented in the first part was based on constructing cross-linked polymers consisting of different number of basic chains and compared with the final results. It was noticed that the more the chains used, the less the scatter of the resulting data compared found though the final trend was not totally as expected.

In the second part the fully cross-linked approach was presented. Compounds randomly move in 3D periodic cell and thus mimic natural movements according to the given boundary conditions. The final results show a clear tendency but not as accurate as could be expected.

It is an unquestionable fact that usage of mesoscale for simulations saves time and CPU power. For this reason for both approaches mesoscale models were introduced. The influence of molecule division per bead was studied for final results and it was noticed that functional groups of the compound should not be split. It was also noted that final results for the mesoscale model are usually less scattered than for the atomistic simulations but the tendency is weaker in this case.

The final conclusion is that a more profound analyzes including a more advanced cross-linking procedure with more accurate simulation parameters is needed, which is planned as a next step of the presented research. The final goal is to predict through the numerical analysis glass transition and coefficient of thermal expansion of the typical cross-linked polymers used in microelectronics packaging.

**Acknowledgments** The authors would like to express their appreciation to Ole Hölck for his valuable advice and discussion in this study. This work was performed in a frame of the "Nanoelectronics for Safe, Fuel Efficient and Environment Friendly Automotive Solutions (SE2A)" project; ENIAC proposal no. 12009. The authors acknowledge Wroclaw Centre for Networking and Supercomputing (WCSS) for the use of modeling software and hardware. Authors acknowledge our coauthors and colleagues that contributed to the publications [3, 6] for useful discussions and valuable remarks.

# References

1. Alder BJ, Wainwright TE (1957) Phase transition for a hard sphere system. J Chem Phys 27:1208
2. Fan HB, Yuen MMF (2007) Material properties of the cross-linked epoxy resin compound predicted by molecular dynamics simulation. Polymer 48:2174–2178
3. Tesarski SJ, Hölck O, Wymysłowski A (2010) Numerical approach to multiscale evaluation and analysis of Tg of cross-linked polymers. In: Conference on thermal, mechanical and multi-physics simulation and experiments in microelectronics and microsystems, EuroSimE, Bordeaux, France
4. Wu CF, Xu WJ (2006) Atomistic molecular modelling of cross-linked epoxy resin. Polymer 47:6004–6009
5. Yarovsky I, Evans E (2002) Computer simulation of structure and properties of cross-linked polymers: application to epoxy resin. Polymer 43:963–9697
6. Wunderle B, Dermitzaki E, Hölck O, Bauer J, Walter H, Shaik O, Rätzke K (2010) Molecular dynamics approach to structure–property correlation in epoxy resins for thermo-mechanical lifetime modeling. Microelectron Reliab (in press)
7. Gee RH, Maiti A, Fried LE (2007) Mesoscale modeling of irreversible volume growth in powders of anisotropic crystals. Appl Phys Lett 90:254105
8. Iwamoto N (2008) Working in relativity in material application development through the use of molecular modeling. Invited paper no. 513107, Fall MRS

# Chapter 14
# Investigation of Coarse-Grained Mesoscale Molecular Models for Mechanical Properties Simulation, as Parameterized Through Molecular Modeling

Nancy Iwamoto

This paper is based upon "Mechanical Properties of an Epoxy Modeled Using Particle Dynamics, as Parameterized through Molecular Modeling" by Nancy Iwamoto which appeared in the proceedings of Eurosime © 2010, IEEE.

**Abstract** The current paper applies both molecular modeling and mesoscale modeling to determine modulus and defect formation in epoxy and epoxy-copper interfaces. The results will show that molecular modeling may be applied directly to parameterize the bead properties used in the mesoscale model, which scales to the physical properties. By parameterizing the mesoscale bead based upon repeat units, it was shown that scaling can be achieved in a larger step than parameterization based upon the typical functional group bead.

## 14.1 Introduction

The linking of scales from the atomistic to bulk finite element models has been a global goal of modeling for many years. One of the conceptual methods used for such scaling is particle dynamics, which is attractive from the standpoint that discrete elements (the "particles") which follow the classic laws of motion are used to bridge the molecular level and the intermediate mesoscale, and from which material properties may be derived for introduction into contium models. Such mesoscale methods to link the scales are available, but to successfully link

N. Iwamoto (✉)
Honeywell Specialty Materials, Moffett Park Drive 1349,
Sunnyvale, CA 94089, USA
e-mail: nancy.iwamoto@honeywell.com

N. Iwamoto et al. (eds.), *Molecular Modeling and Multiscaling Issues*
*for Electronic Material Applications*, DOI: 10.1007/978-1-4614-1728-6_14,
© Springer Science+Business Media, LLC 2012

the scales parameterizations should be introduced to represent the correct chemistry of the material. Parameterizations to make the methods generally useful are sparse. Particle dynamics, as practiced in Mesocite within the Materials Studio graphic environment (from Accelrys, Inc.), allows individual tuning of the particle interactions so that parameterization tuning becomes easy to introduce.

The philosophy employed to parameterize the mesoscale models has been derived from observations using molecular modeling for property trends in other work [1–6], and is based upon the concept that all properties are derived from a series of interactions. The root of the interaction should be obtainable from the molecular scale, by calculating the interaction of each important interface or each important interaction. For instance, it has been found by previous examination that molecular scale models represent the adhesive modulus adequately enough to impact formulation work. And, in cases where the formulations are carefully represented, the modulus is quite accurate.

Although classically parameterized coarse-grained molecular models [7, 8] usually examine smaller functional group units from which they define and parameterize the particles (or "beads"). The current parameterization philosophy was directly extrapolated from previous work looking at parameter development for silica particle-underfill interactions using discrete element modeling (DEM) which concluded that averaging across more atoms might be possible [9–11]. In that case, it was learned that if the silica particle interactions could be parameterized from molecular models of approaching silica surfaces with the binder molecules placed in-between, the energy changes occurring at each step could then describe energy/separation curves and forces could be established that represent the filler particle interaction. These energy curves were used to parameterize the particle models and then the particles discretely modeled using DEM (discrete element modeling) [9–11], and an accurate accounting of how the particles moved during the underfill process was obtained. Parameterizations based upon the type of binder taught us how to change the formulation, and improve the underfill characteristics, which further taught us that a jump in length scales using discrete models could be obtained in order to achieve accurate silica particle dynamics.

This work was designed to incorporate what was learned about larger silica particle parameterizations and apply this to the coarse-grained mesoscale molecular models in order to investigate the use of larger bead representations of large polymer repeat units. The work includes investigation of parameterization of the large repeat units to achieve a jump in scale from what is normally used in coarse-grained molecular models and the resulting simulations to look at the mechanical properties using a simple tensile deformation. Since any scale-jump may be made considering the relevant interactions present from the molecular levels and which entities work as groups, the large bead size is appropriate. This is especially true in the current work where relatively rigid repeat units are represented.

## 14.2  Background: Molecular Models and the Derived Mesoscale Parameters

The molecular models used in this paper represented an epoxy system which was characterized by TUDelft, comprising an epoxy novolac cured with bisphenol A [12] in studies of epoxy molding compounds. Both the epoxy cohesive interface as well as the interface between the epoxy and copper(I)oxide was modeled. All molecular modeling representing the bulk (unfilled) polymer and the copper oxide-polymer interfaces (representing leadframe interfaces) used the CVFF force field supplied within the InsightII/Discover software from Accelrys [13]. In a similar procedure previously used to determine the modulus of dielectric polymer materials and adhesion in die attach adhesives [2–6], molecular models were used to determine parameters for the coarse-grained bead model. The molecular model was first reduced to the most likely repeat units, with the size of these repeat units used to represent the particle to particle or "bead" to "bead" bond size for the mesoscale model. The repeat unit interaction energies (to be used for the mesoscale model) were then derived with procedures used to locate modulus during molecular dynamics (MD) [1–6], in which the small repeat units (MW $\sim$ 800) which were previously minimized together were given a series of increasing forcing potentials to shear them apart. For each run, the shearing action was allowed to proceed until the molecules were separated, with the energies proceeding thru a maximum. The initial energy response slope was monitored until a deviation in linear response with speed was found, and the trajectory from this run was used to find the maximum energy change. The maximum energy change upon separation was used for the interaction energies (the VDW, or Van der Waals energies) and the average separation distances between the repeat units from a larger 20 unit oligomer construction used to estimate the bead bond distances. In a similar manner, the copper (I) oxide adhesive interface was modeled using the same epoxy repeat unit optimized on a single crystal $Cu_2O$, copper(I)oxide layer ([100] surface) which had been optimized on a fixed copper [100] surface. Since the parameterizations used were modified from the default Martini Force Field supplied within the Accelrys software, an additional improvement would be the determination of the force constants. For the current work the default force constants were used. As may be expected additional perfection is warranted for the parameterizations.

The initial "bead" or "coarse-grained" mesoscale models were then generated from those molecular models derived from the oligomer of the epoxy repeat units and the crystal surface created for $Cu_2O$. The structures were generated within 3D periodic unit cells to reproduce a polymer density target of 1, sitting next to an oxygen terminated $Cu_2O$ layer generated from the [100] cut of the $Cu_2O$ unit cell (for $Cu_2O$, the unit cell was supplied from the Accelrys database). The bead coarse graining was done automatically within the Mesocite software supplied by Accelrys, from the previously defined $Cu_2O$ beads [13]. For the copper oxide, a $2 \times 2$ $Cu_2O$ crystal supercell unit (available in the Materials Studio database) was used as a bead, with the cell length used as the bond stretch. Replication and supercell generation

**Table 14.1** Modified
parameters used in mesocite
(Edited Martini Force Field)

| | VDW (kcal/mole) | Bond Stretch R0 (A) |
|---|---|---|
| Epoxy repeat unit | 55 | 24.77 |
| Epoxy-Cu$_2$O | 64 | 2.1 |
| Cu$_2$O(Cu)—oxygen terminated surface | 90 | 2.1 |

were used to grow the polymers and surface to the large size blocks. The large cell blocks were energy optimized and equilibrated (constant volume and temperature conditions) to room temperature before deformation was applied to simulate stress. (Free-standing models were also done, but will not be discussed in this paper.) To our knowledge this is the first example of a large jump in bead size performed for determination of mechanical response of a polymer and polymer-metal interface and generating the interfacial stress–strain curves. Both full periodic unit cells (no vacuum layer) and vacuum unit cells were generated and compared. In all full periodic cell cases, the size of each side of the polymer block obtained was around 500 A (50 nm), with the approximated molecular weight inside the periodic cell >78 M. The vacuum unit cells were initially constructed from the full periodic cell models, but final models were reduced in size once the basic deformation procedures had been established in the full periodic cells.

The parameters used for the mesoscale modeling are found in Table 14.1 and the structures and the final large-block periodic cell for the uncrosslinked and cross-linked polymer is represented in Fig. 14.1.

## 14.3 Results of Deformation Tests

In the current cases reported, periodic cells were used to represent the bulk, and deformation of the cells was used to obtain the stress-state assumed to be present during tensile testing. Both full cell and vacuum cell cases were tested. For each case, the unit cell was stressed along one dimension (vertial direction in all figures), and the other two dimensions adjusted to keep the total density constant; so for each tensile increment, the periodic cell was rebuilt with the new dimensions. Although constant density is recognizably not a normal response, it was considered to be reasonable for a first evaluation of these large bead models until methods could be evalutated for more realistic deformation to take into account volume changes [14, 15]. In addition, as will be discussed at the end of the adhesive case section, vacuum cells were also constructed in which only the copper oxide was moved and constant volume of the material mass was not considered in order to isolate the adhesive interface response from contributions of the rest of the mass in the unit cell.

For each step increment in deformation, energies were obtained in-line with the philosophy that each step represents a non-equilibrium state (as opposed to an

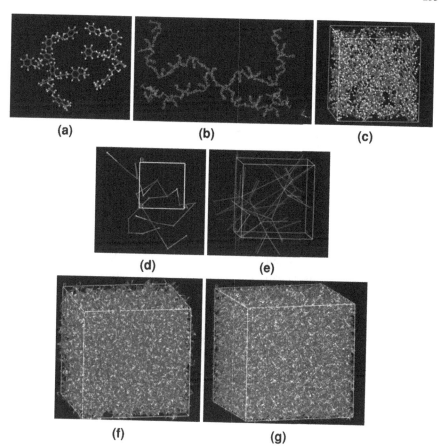

**Fig. 14.1** Model examples. **a** Atomistic repeat units used for coarse-grained beads; **b** atomistic oligomer from repeat units; **c** periodic cell of oligomer; **d** coarse grain of oligomer ("uncrosslinked case") unit cell; **e** fully cross-linked and coarse-grained unit cell with same size oligomer; **f** energy minimized supercell of uncrosslinked case; **g** energy minimized supercell of cross-linked case

equilibrium process which would require long dynamic runs and would be more associated with stress relaxation over time). The energy of the structure was first relaxed using a minimum number of steps to obtain the start of an energy plateau. The structure was thus optimized enough to allow temperature equilibration in a second step. In the cases reported it was found that just 100 steps of NVT (constant number of particles, volume and temperature) minimization was required. The periodic cell was then given a minimal equilibration at room temperature using constant volume and constant temperature dynamics (again using a minimum number of steps, with the assumption that too many equilibration steps may unrealistically relax out the strain. In the cases reported, it was found that the minimum of 0.1 ps with 1 fs steps was adequate). The energy obtained after equilibration was used as the system energy in an energy/strain curve. The initial slopes were used to estimate modulus and the estimated yield was taken at the maximum point in the energy curve.

Other deformation types used included the use of a vacuum cell and displacement of only the top and/or bottom layer(s) of atoms and use of an optimized non-periodic cell and displacement of only the top and/or bottom layer(s) of atoms including the use of a rough $Cu_2O$-epoxy interface. These cases will be discussed in later work [14, 15], but are mentioned here for completeness.

### 14.3.1 Bulk Cohesive Cases

The bulk cases started from molecular models of an uncrosslinked form (each strand a connected double T shape of molecular weight around 23 K), and a cross-linked case in which the strands were linked in all three (X, Y, Z) dimensions within its unit cell and then coarse grained to beads. These cases served as a sanity check to be sure protocols were appropriate for the copper-epoxy interface modeling. Three basic reality checks were monitored: (1) appropriate magnitudes of the estimated modulus; (2) lower expected modulus of the uncrosslinked versus cross-linked forms; (3) reasonable yield points (2–4%). It was also hoped that the onset of defect formation could be observed in these models, to explain failure of the polymer. However, as mentioned previously, since the entire parameterization set was not perfect, only estimates and appropriate trends were expected.

Examples of the energy "stress–strain" curve obtained from the stepwise deformation are found plotted with pictures of the models in Figs. 14.2, 14.3, 14.4, 14.5 and 14.6. Table 14.2 contains a tabular form of all of the model results. The modulus result was directly obtained from the initial slope of the derived stress/strain curve. The raw result, the energy/distance slope in kcal/mole/A was then converted to the equivalent energy J/mole/A and then to nt/mole (considering that each model result represents $6 \times 10^{23}$ unit cells), and then further converted to Pa (n/m$^2$) using the specific cross-section of the unit cell. The results are reported in GPa units. As may be noticed, the modulus of the uncrosslinked system is much lower than the modulus of the cross-linked system, using the same mesoscale parameters to generate the simulation. The magnitudes of the moduli are reasonable and comparable to the measured values for the cured epoxy [12]. The expected yield ranges are also reasonable (around 3%) and the estimated yield appears to occur later for the uncrosslinked system than the cross-linked one, which may show the effect of higher chain mobility in the uncrosslinked case. However more work must be done to understand the energy response past the elastic range in these models.

Interestingly, the molecular model shows similar modulus to the cross-linked system, which is reasonable given that the molecular models were constructed with maximum interaction that might be present in a cross-linked case. The cohesive mesoscale modulus (Table 14.2) also agrees with the findings of previous

**Fig. 14.2** Void formation in mesoscale periodic full cell models. **a** Side view of the uncrosslinked epoxy before deformation and **b** after deformation

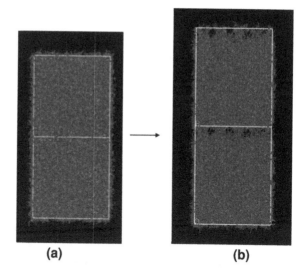

(a)                                      (b)

molecular models of modulus of cross-linked structure [2], and justifies that small non-periodic molecular models can reproduce the modulus when the initial resistance to stress is derived from a small initial deformation on the molecular scale. More importantly the cohesive modulus served to benchmark the mesoscale method to experimental [12]; and with the same mesoscale cohesive modulus repeated from the atomistic cohesive modulus served to demonstrate a continuity of scale. In addition, the expected increase in modulus with cross-linking is found.

One of the most significant observations found in these calculations is that void growth (growing from voids existing from the original optimized polymer) is clearly found. Void growth is found in both the uncrosslinked and the cross-linked models, with observed onset occurring at approximately similar strain. The voids are most easily found for the uncrosslinked case. Figure 14.2 shows the voids before and after deformation, with the energy trajectory found in Fig. 14.3. Figure 14.3 also shows a horizontal slice through the void region before and after deformation clearly showing more voids and higher growth after yield.

Views for the cross-linked case are shown in Figs. 14.4, 14.5 and 14.6. These structures are highly cross-linked and void formation is not as radical as the uncrosslinked case, but is most noticeable in the close up side view (Fig. 14.4d). It is also clear that there is material thinning occuring, as viewed in the vertical cross-sectional slice in the middle (Fig. 14.5) where there appears to be less material.

Figure 14.6 shows the energy response along with the slice figures at yield for the cross-linked case, analogous to the uncrosslinked case. A small void is found to develop at the energy maximum (Figs. 14.5, 14.6). Clearly there is no large growth of void regions in the cross-linked case as found in the uncrosslinked case. However, there is found to be void growth just after the maximum in energy is obtained, as shown in Fig. 14.5, with an expanded sideview in Fig. 14.6.

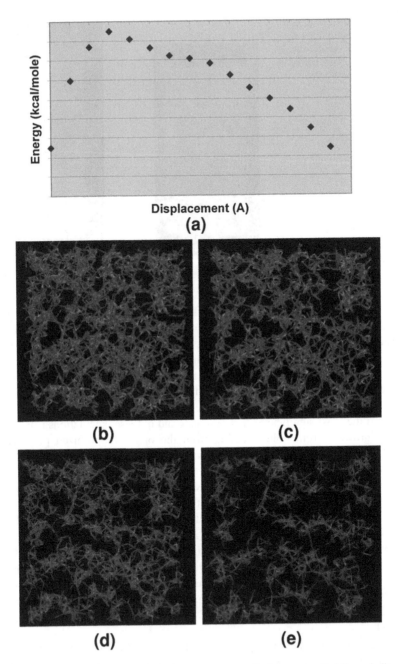

**Fig. 14.3 a** Example of uncrosslinked epoxy energy-displacement of full periodic cell (Fig. 14.2) with accompanying horizontal cross-sections showings void at **b** start **c** energy peak **d** one step after energy peak and **e** at trajectory end (same as after deformation model in Fig. 14.2). Void growth is evident, accelerating after the energy peak

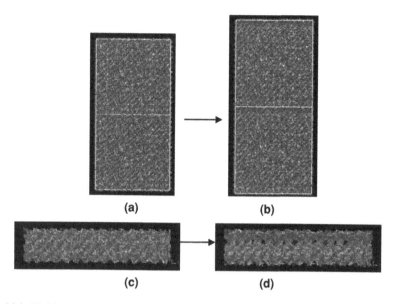

**Fig. 14.4** Highly cross-linked case. **a** Before deformation, **b** after deformation, **c** close up of the thinned region before and **d** after deformation. Clearly void growth is found

**Fig. 14.5** Highly cross-linked model, horizontal slices extracted from models. **a** At start of trajectory and **b** at last step showing void concentration in middle

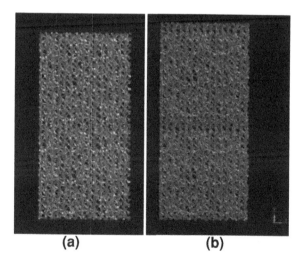

## 14.3.2 Adhesive Copper-Epoxy Interface Cases

As with the bulk case, the adhesive models were built around periodic cells. However, because of the limitations of the computer used (which limited the number of beads that were practical to calculate in a reasonable time), a very thin monocrystalline layer of $Cu_2O$ was used for the initial coarse-grained cell.

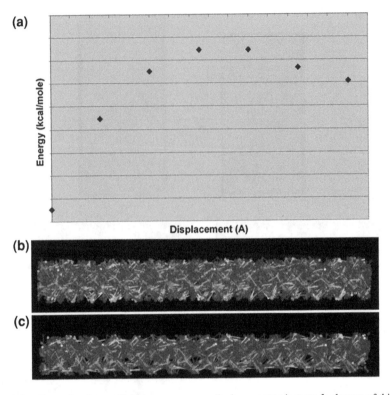

**Fig. 14.6** **a** Example of cross-linked epoxy energy-displacement trajectory, **b** closeup of thinned area at energy max, **c** closeup of thinned area one step after energy max, showing void response

**Table 14.2** Properties of epoxy cohesive models

|  | Modulus (Gpa) | % elongation at yield | Void growth onset (% elongation) |
|---|---|---|---|
| Experimental cohesive | ~2 | – | – |
| Repeat unit (molecular) | 2.0 | – | – |
| Uncrosslinked full periodic cell (mesoscale) | 0.97 | 2.9 | ~3.8 |
| Cross-linked full periodi cell (mesoscale) | 2.13 | 3.4 | ~3.9 |

For the adhesive models, the configurations were constructed in two different formats. Vacuum cells (Fig. 14.7) were used so that a specific $Cu_2O$-interface separation could be specifically targeted and manipulated. The interface area was then built into a thin full-packed continuous non-vacuum cell (Fig. 14.8) in order to apply cell deformation similar to the cohesive cases (Table 14.3, condition A).

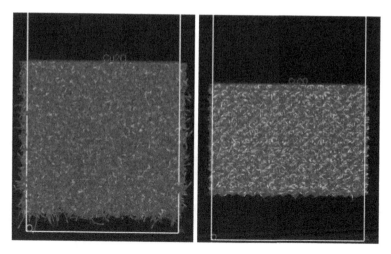

**Fig. 14.7** Cu-epoxy periodic vacuum cells side view (Top monolayer of $Cu_2O$ is barely visible)

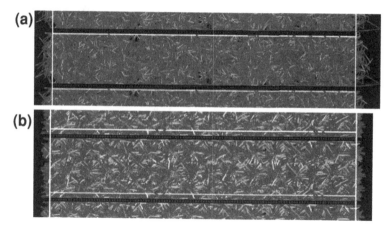

**Fig. 14.8** Full cell (3 periodic cells shown). **a** Uncrosslinked. **b** Cross-linked

For the vacuum cells, the $Cu_2O$-epoxy adhesion was tested in two ways. The first was a simple incremental stepping of the $Cu_2O$ layer into the vacuum space in an attempt to isolate the $Cu_2O$-epoxy atomic interface (Table 14.3, condition B), with a thin layer of bottom beads (the least number of beads representing a continuous bead layer at the bottom of the unit cell) constrained to avoid possible creep of the entire polymer mass upward with the $Cu_2O$ movement. The second type of model used both the step increment of the $Cu_2O$ layer into the vacuum as well as cell deformation to keep the material volume constant (Table 14.3, condition C). Condition C was intended to include contributions from the bulk of the epoxy. For all the cases, the models were equilibrated to RT, before

**Table 14.3** Properties derived from epoxy-Cu adhesive models

|                            | Modulus (Gpa) | % elongation at yield (at high E) |
|----------------------------|---------------|-----------------------------------|
| Uncrosslinked-condition A  | 4.87          | 4.5                               |
| Uncrosslinked-condition B  | 5.01          | 2.0                               |
| Uncrosslinked-condition C  | 6.26          | 2.4                               |
| Cross-linked-condition A   | 7.07          | 4.0                               |
| Cross-linked-condition B   | 6.76          | 2.4                               |
| Cross-linked-condition C   | 12.63         | 4.4                               |

Condition A: full-packed cell deformation; condition B: vacuum cell, movement of $Cu_2O$ into vacuum; condition C: vacuum cell, movement of $Cu_2O$ layer + cell deformation

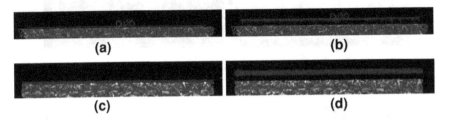

**Fig. 14.9** Close up side view of vacuum cell adhesive interface region of Fig. 14.7. **a** and **b** are the uncrosslinked cases and **c** and **d** are the cross-linked cases. Before deformation is shown in **a** and **c** and after deformation is shown in **b** and **d**. Both cases show clean polymer separation from the copper oxide

deformation, keeping the $Cu_2O$ beads fixed in their crystalline configuration. Upon deformation, the constraints on the $Cu_2O$ were removed. Both the uncrosslinked and cross-linked epoxy cases were modeled for the $Cu_2O$ interface. Table 14.3 contains a tabular form of all of the model results and examples of the resulting four cases are found in Figs. 14.8, 14.9, 14.10, 14.11 and 14.12.

For all of these adhesive cases clean delamination was found, with no apparent epoxy attraction to the $Cu_2O$ surface during the deformation steps. Even for a full periodic case, both the uncrosslinked (Figs. 14.9a, b and 14.11a, b) and cross-linked cases (Figs. 14.10c, d and 14.12c, d) did not show the large void growth as found in the cohesive models. Although the clean delamination does not adequately portray the experimental results, the energy trends suggested that cohesive failure should occur before adhesive failure. That is, the adhesive moduli were found to be much higher than the cohesive moduli and the cross-linked adhesion was found to be higher than the uncrosslinked case. These comparisons suggested that when more of the epoxy is included into the response by deforming the periodic cell (Table 14.3, condition C), the adhesive modulus found is generally higher than the condition in which the epoxy/$Cu_2O$ interface deformation was isolated (condition B). In addition conditions A and C had higher yield than condition A indicating that there is contribution from the epoxy. Use of a cell deformation (for conditions A and C) was intended to better match conditions

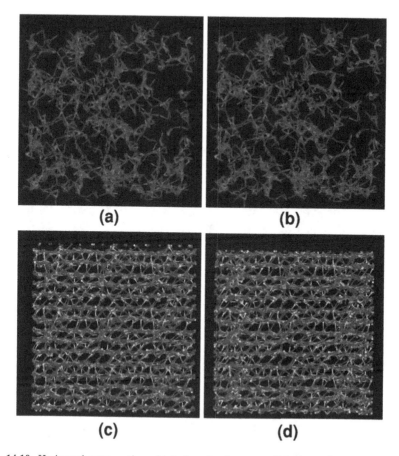

**Fig. 14.10** Horizontal cross-section of interface for the uncrosslinked case in **a** and **b** and the cross-linked case **c** and **d**, where **a** and **c** are before deformation and **b** and **d** are after deformation. Unlike the cohesive case there is little or no void growth and no change in the polymer indicating adhesive failure

used in the full cell cohesive models, but the adhesive results tended to underscore the importance of understanding material coupling (or both the adhesive to cohesive coupling as well as the extent of cohesive coupling into the bulk) as different adhesive modulus was attained depending upon condition and model configuration.

Since, clean delamination for this epoxy is not reasonable at maximum experimental process conditions, better isolation of effects was necessary, and conditions were investigated to help improve material coupling in the models. For instance, one reason for the failure of the correct adhesive simulation is probably due to failure of the software to adequately treat bond rupture. So, a bead bond rupture criterion was determined from molecular modeling by deforming a single repeat unit to be used at each step. The maximum length obtained before

**Fig. 14.11** Side view of a thin full periodic cell adhesive model, uncrosslinked case **a** before and **b** after deformation; and cross-linked case **c** before and **d** after deformation (after achieving an energy maximum) Neither cases indicate void formation

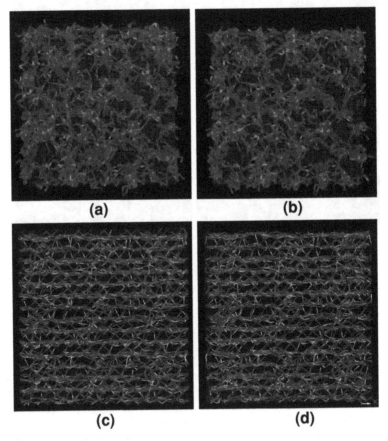

**Fig. 14.12** Horizontal cross-section at interface of a thin full periodic cell adhesive model, uncrosslinked case **a** before and **b** after deformation; and cross-linked case **c** before and **d** after deformation (after achieving an energy maximum). Unlike the cohesive cases, there is no voiding

unnatural bond elongation and energy loss was determined in molecular models. In the case of the current repeat unit, a 34 A bead bond rupture criterion was found. However, another cause of the unrealistically clean delamination found in

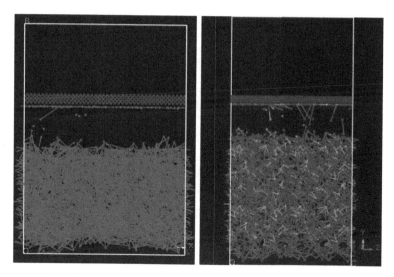

**Fig. 14.13** Effect of including bead bond rupture criterion at each step. *Left* fully cross-linked, adhesive modulus 2.9 GPa. *Right* uncrosslinked adhesive modulus 0.4 GPa

the above models could also be due to inadequate definition of the polymer density involved at the $Cu_2O$ interface, and optimization procedures were implemented.

So, in order to address both above concerns, a new vacuum cell was constructed. Figure 14.13 shows the result of introduction of a bond rupture criterion for the adhesive case using a vacuum cell case in which the copper oxide layer was fully optimized on top of a cross-linked epoxy polymer, with the bottom bead layers of the epoxy fixed as before to avoid translation of the entire mass. To build the interfaces and ensure that the entire mass (copper oxide + epoxy) was at a global miminm, cycles of heating, compression and cooling were used in which the copper oxide was slowly pushed into the epoxy at 1 A steps, the system was equilibrated at room temperature, heated to above 500 K and cooled again to room temperature (RT). As expected, the model energies began to fall, initially with each compression step indicating that there is inadequate density definition at the $Cu_2O$ interface. The steps were repeated until the RT energies of the unit cell began to increase. These steps were thought to be a fair representation of experimental changes at the interface especially since experimental compression is usually done when the polymer is at a less viscous state. In this manner both compression and heating could help simulate the flow changes at the interface. As a vacuum cell was used, the unit cell was kept constant at each step, and only the copper oxide layer was manipulated. For the tensile deformation steps, the copper oxide was then moved away from the epoxy at 2 Å increments per step. At each step, the unit cell was equilibrated to room temperature for only 1 ps, long enough to just reach an energy plateau. (As before the rationale is to attempt to representate a dynamic process.) After each step, the bonds were analyzed for lengths above the bond rupture criterion, and if found the bonds were broken. The procedure was repeated for three different levels of cross-linking.

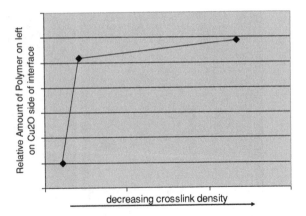

**Fig. 14.14** Effect of cross-link density on the amount of polymer left behind on the $Cu_2O$ side of the interface after total interfacial failure

As may be observed by Fig. 14.13, this new treatment resulted in a more realistic adhesive interface delamination in which some beads were left behind on the $Cu_2O$ side. As expected the uncrosslinked case left a larger residue on the copper oxide side of the interface than the cross-linked case (Fig. 14.14). Full stress–strain curves could also be extracted from the energy versus deformation data [14, 15] which were used to show the impact of cross-link density. The moduli extracted from these cases showed similar magnitudes with approximately 2.9 GPa for the fully cross-linked case and approximately 0.4 GPa for the uncrosslinked case.

These new vacuum cell constructions could also be used for cohesive models, by moving the top half of the model up into the vacuum. The same bead bond rupture was used for the cohesive models indicating lower modulus for the cohesive case than the adhesive case with the cross-link modulus at approximately 2GPa, consistent with the experimental cohesive modulus [12] as well as the predicted modulus derived from both the previous full cell mesoscale models and molecular models. The study of bead bond criterion will be treated in future papers [14, 15] as a means to develop the full stress strain curve to failure and to determine effects of the crosslink density and void formation.

## 14.4 Conclusions

In the current work, initial parameterizations for use with a phenolic epoxy adhesive have been derived from the molecular scale and models were scaled from a molecular modeling 300-atom level to a particle scale which represents over 70 million atoms, using a standalone workstation. The starting parameters, as determined by molecular modeling were used to derive properties from the mesoscale level models. In addition, the molecular model shows similar modulus to the cross-linked system, which is reasonable given that the molecular models assume cross-link structure. This observation helps to establish the molecular link to bulk properties

The discrete model behavior so far matches expected bulk behavior, including cohesive modulus (showing cohesive modulus continuity between the molecular scale and mesoscale, through to verified experimental modulus), and void formation at energy trends which indicate yield. Void generation is best illustrated in the uncrosslinked case in which a more random distribution of polymer is achieved after optimization than the crosslinked case, and voids were found to grow into one another. The cross-linked case still has remnants of the repeat structure due to the supercell origins. However in both cases it appears that voids are growing and in both cases, the energy maximum (used as yield) may be associated with observable void growth. The issue of material coupling between the adhesive and cohesive interface still remains an important topic to fully understand energy distribution during stress, but the introduction of a bead bond rupture/failure was especially useful to obtain realistic adhesive failure using a vaccum cell, and demonstrated varying amounts of cohesive failure contribution depending upon the crosslink density. Defining the material response coupling between the adhesive interface and cohesive interface still remains an important topic.

There is more work that must be accomplished with the parameterizations, but initial work shows that parameterization and scaling to large monomer beads containing different functional groups is feasible (and quick) using an energy interaction derived from the molecular level. In addition, the initial work with bead bond failure criterion shows that this is an important variable to impart reality into the adhesive bond model and should improve the adhesive interface prediction. However most significantly, because mesoscale models are derived directly from the molecular level and so retain the major interaction energies and architecture of the molecular bonding, the demonstration of scale continuity suggests that this method can be a good way to scale and still maintain tracking back to the underlying molecular contribution.

Future work needed includes bead refinements for the force constants, but it is suggested that other refinements could also come from bead shape parameterization, as previous studies with discrete element modeling suggests that shape factors greatly improve the success of the prediction [10]. As mentioned previously, understanding the proper coupling between the adhesive and cohesive interfaces needs work. Toward this end determination of the effects of interfacial density and roughness effects are underway.

In addition, rate correlations to the mesoscale must be made in order to understand the computational time domains represented. So far, rate trends have been noticed, but quantitative confirmation of rate scaling using mesoscale models should be done.

**Acknowledgments** Software is provided by Accelrys, Inc. (San Diego, CA) funded from the Seventh Framework Program for Research and Technological Development (FP7) of the European Union (NMP3-SL-20080214371) in the Nanosciences, Nanotechnologies, Materials and New Production Technologies Program.

# References

1. Iwamoto N (2008) Working in relativity in material application development through the use of molecular modeling, invited paper #513107, Fall MRS, Boston, MA December, pp 1–5
2. Iwamoto NE, Moro L, Bedwell B (2002) Understanding modulus trends in ultra low k dielectric materials through the use of molecular modeling. In: Proceedings of the 52nd electronic components and technology conference, 28–31 May 2002, San Diego, CA, pp 1318–1322
3. Iwamoto N, Lee E, Truong N (2004) New metal layers for integrated circuit manufacture: experimental and modeling studies. Thin Solid Films 469–470:431–437
4. Iwamoto N, Pedigo J, Grieve A, Li M (1998) Molecular modeling as a tool for adhesive performance understanding. In: Proceedings of the MRS 98 symposium J: electronic packaging materials science X, vol 515, San Francisco, CA, pp 23–30
5. Iwamoto N, Pedigo J (1998) Property trend analysis and simulations of adhesive formulation effects in the microelectronics packaging industry using molecular modeling. In: Proceedings of the 48th electronic components and technology conference, 25–28 May, pp 1241–1246
6. Iwamoto NE (2000) Advancing materials using interfacial process and reliability simulations on the molecular level. In: Proceedings of the 5th international symposium and exhibition on advanced packaging materials, March 2000, Braselton, GA, pp 14–17
7. Nielsen SO, Lopez CF, Srinivas G, Klein ML (2004) Coarse grain models and the computer simulation of soft materials. J Phys Condens Mater 16:R481–R512
8. Siewart J, Marrink SJ, Risselada HJ, Yefimox S, Tielman DP, de Vries AH (2007) The Martini force field: coarse grained model for biomolecular simulations. J Phys Chem B 2111: 7812–7824
9. Iwamoto N, Nakagawa M, Mustoe GGW (1999) Simulation underfill flow for microelectronics packaging. In: Proceedings of the 49th electronic components and technology conference, 1–4 June, San Diego, CA, pp 294–301
10. Iwamoto N, Li M, McCaffrey SJ, Nakagawa M, Mustoe G (1998) Molecular dynamics and discrete element modeling studies of underfill. In: 31st international symposium on microelectronics, San Diego, CA, 1–4 Nov 1998. Also published in 1998 international journal of microcircuits and electronic packaging, vol 21(4), fourth quarter 1998, pp 322–328
11. Mustoe GGW, Nakagawa Lin MX, Iwamoto N (1999) Simulation of particle compaction for conductive adhesives using discrete element modeling. In: Proceedings of the 49th electronic components and technology conference, San Diego, CA, pp 353–359
12. Jansen K, testing results obtained from the NanoInterface Consortium funded from the Seventh Framework Program for Research and Technological Developmen (FP7) of the European Union (NMP3-SL-20080214371) in the Nanosciences, Nanotechnologies, Materials and New Production Technologies Program
13. Molecular Modeling software used was Discover and InsightII. Mesoscale modeling software used was Mesocite. Both are from Accelrys (San Diego, CA)
14. Iwamoto N (2011) The use of mesoscale modeling to understand polymer failure in electronic packaging. Presented at international plasticity 2011, 3–8 January, Puerto Vallarta, Mexico
15. Iwamoto N (2011) Developing the mesoscale stress-strain curve to failure. In: Proceedings of Eurosime, Linz, Austria, 18–20 April 2011

# Disclaimer

Although all statements and information contained herein are believed to be accurate and reliable, they are presented without guarantee or warranty of any kind, express or implied. Information provided herein does not relieve the user from the responsibility of carrying out its own tests and experiments, and the user assumes all risks and liability for use of the information and results obtained. Statements or suggestions concerning the use of materials and processes are made without representation or warranty that any such use is free of patent infringement and are not recommendations to infringe any patent. The user should not assume that all toxicity data and safety measures are indicated herein or that other measures may not be required.

# Abbreviations

| | |
|---|---|
| **3DPC** | 3D periodic cells |
| **AFM** | Atomic force microscopy |
| **BARC** | Bottom anti-reflcoating |
| **BPA** | Bisphenol A |
| **BPSG** | Boron phosphate silicate glass |
| **CAD** | Computer aided design |
| **CFD** | Computational fluid dynamics |
| **CGMD** | Coarse grained molecular dynamics |
| **CME** | Coefficient of moisture expansion |
| **CNA** | Common neighbor analysis |
| **CNT** | Carbon nanotube |
| **COMPASS** | Condensed-phase optimized molecular potentials for atomic simulations studies force field |
| **CTE** | Coefficient of thermal expansion |
| **$Cu_2O$** | Copper (I) oxide |
| **CVFF** | Consistent valence force field |
| **CZM** | Cohesive zone model |
| **D** | moisture diffusion coefficient |
| **DEM** | Discrete element modeling |
| **DFT** | Density functional theory |

N. Iwamoto et al. (eds.), *Molecular Modeling and Multiscaling Issues for Electronic Material Applications*, DOI: 10.1007/978-1-4614-1728-6, © Springer Science+Business Media, LLC 2012

| **DGEBA** | Diglycidylether of Bisphenol A |
| **DMA** | Dynamic mechanical analysis |
| **EAM** | Embedded atom method |
| **ECAE** | Equi-channel angular extrusion |
| **EGBD** | Extrinsic grain boundary dislocation |
| **EMC** | Epoxy molding compound |
| **EMD** | Equilibrium molecular dynamics |
| **FBD** | Field to breakdown |
| **FEA** | Finite element analysis |
| **FEM** | Finite element modelling |
| **FFV** | Fractional free volume |
| **FPD** | Flat panel display |
| **FSG** | Fluorinated silica glass |
| **FTIR** | Fourier transform infrared spectroscopy |
| **G(IC)** | Critical energy release rate |
| **GGA** | Generalized gradient approximation |
| **iBARC** | Inorganic bottom anti-reflective coating |
| **IC** | Integrated circuit/chip |
| **IGBD** | Intrinsic grain boundary dislocation |
| **ILD** | Interlayer dielectric |
| **ITRS** | International technology roadmap for semiconductors |
| **K(eqv)** | Equivalent stress intensive factor |
| **LAMMPS** | Large scale atomic/molecular massively parallel simulator |
| **LB** | Langmuir-Blodgett |
| **LC** | Liquid crystal |
| **LCF** | Low cycle fatigue |
| **LDA** | Local density approximation |
| **m(e)** | Effective mass |
| **MD** | Molecular dynamics |
| **MFP** | Mean free path |

| | |
|---|---|
| **MM** | Molecular mechanics |
| **MWCNT** | Multiwalled carbon nanotube |
| **NEMD** | Nonequilibrium molecular dynamics |
| **NMR** | Nuclear magnetic resonance |
| **NPT** | Constant number of particles, pressure and temperature |
| **NVE** | Constant number of particles, volume and energy |
| **NVT** | Constant number of particles, volume and temperature |
| **oBARC** | Organic bottom anti-reflective coating |
| **OSG** | Organosilicate glass |
| **PALS** | Positronium annihilation lifetimer spectroscopy |
| **PBC** | Periodic bound conditions |
| **PBE** | Perdew, burke, and ernzerhof (a common form of a GGA) |
| **PCFF** | Polymer consistent force field |
| **PSG** | Phosphate silicate glass |
| **QC** | Quasicontiuum method |
| **SAM** | Self assembled monolayer |
| **SOG** | Spin-on-glass |
| **SWCNT** | Single walled carbon nanotube |
| **TBCB** | Tapered double cantilever beam |
| $T_g$ | Glass transition temperature |
| **TIM** | Thermal interface materials |
| **ThOx** | Thermal oxide |
| **UFG** | Ultrafine grained metals |
| **VDW** | Van der Waals |
| **VFTL** | Via first trench last |
| **VHR** | Voltage holding ratio |
| **VIB** | Virtual internal bond |
| **VOF** | Volume of fluid model |

# Index

## A
3D periodic cells, 3DPC, 225
Adhesion, 190, 191, 194, 197, 199
Adhesive to cohesive coupling, 243
Aging, 40, 41, 43, 45, 51
Algorithm, 150–151, 175, 184
ANSYS, 205, 208
Anti-reflective coating, 39
Asymmetrical grain boundaries, 58, 64
Atomic force microscopy, AFM, 199
Atomistic simulation, 1, 229, 172, 55
Azobenzene, 204–206, 209–211

## B
Band gap, 6–7, 15, 18, 28, 29, 32, 35, 36
Bead bond, 233, 244, 246, 247
Bead bond failure criterion, 247
Bead parameterization, 232, 233, 236, 246, 247
Beads, 233, 247
Bisphenol A, BPA, 135
Boron phosphate silicate glass, BPSG, 232
Bottom anti-reflective coating, BARC, 39, 40
Boundary conditions, 118, 120
Brittle Fracture, 11

## C
Capacitance, 26, 36
Carbon nanotube, CNT, 42
Chemical properties, 4
Chip to package copper interconnect, 41
CNT chirality, 108
CNT diameter and thermal conductivity, 93

CNT length and thermal conductivity, 93
Coarse grain, 5
Coarse grained molecular dynamics, CGMD, 150
Coarse-grained mesoscale, 150
Coating, 203, 209
Coefficient of moisture expansion, CME, 12
Coefficient of thermal expansion, CTE, 11, 13
Cohesion, 236
Cohesive coupling into the bulk, 243
Cohesive elements, 191, 195, 197
Cohesive zone, 189, 191, 194, 197–199
Cohesive zone model, CZM, 194
Common neighbor analysis, CNA, 58
Complementary metal-oxide-semiconductors, CMOS, 17
Compression, 93, 95–97
Computational fluid dynamics, CFD, 205
Computer aided design, CAD, 214
Concurrent multiscale method, 55, 56
Condensed-phase Optimized Molecular Potentials for Atomic Simulations Studies force field, COMPASS, 107, 135
Consistent valence force field, CVFF, 233
Constant number of particles, volume and energy, NVE, 96
Constant number of particles, volume and temperature, NVT, 96, 107
Constitutive, 190, 191–192, 194, 197, 199
Constitutive cohesive relation, 194, 197
Constitutive law, 190
Contact angle, 203–205, 207–211
Contact angles, 204, 208, 210
Copper (I) oxide, $Cu_2O$, 233

N. Iwamoto et al. (eds.), *Molecular Modeling and Multiscaling Issues for Electronic Material Applications*, DOI: 10.1007/978-1-4614-1728-6,
© Springer Science+Business Media, LLC 2012

**C** *(cont.)*
Copper interconnect, 87
Correlating structure and property
Coupling agent, 133–134, 142
Crack growth, 77–79, 81, 85, 87, 88–89
Crack growth analysis, 78, 87, 89
Critical energy release rate, G(IC), 138
Crosslinked, 239
Crosslinked epoxy resin, 151
Cross-linked polymers, 213–214, 229
Crosslinking algorithm, 165, 172–173, 229

**D**
Defect formation, 231, 236
Delamination, 189, 191, 194–195, 197–199
Density, 155–157, 162, 166, 169, 170, 176,
        180–181, 208, 210, 213–214, 217,
        219, 222, 225–227, 229
Density functional theory, DFT, 4, 5
Die attach adhesives, 233
Dielectric, 25–28, 33, 34, 46
Dielectric constant, 15
Dielectric properties, 3, 15–16, 21
Dielectric susceptibility, 25
Diffusion, 3, 13, 21, 63, 133, 134
Diffusion coefficient, 149, 155, 157–158, 167,
        171, 180
Diffusive, 103, 111
Diglycidylether of bisphenol A, DGEBA, 196
Discrete element modeling, DEM, 232
Discrete elements, 231, 247
Dislocation, 55–56, 63–65, 68, 70–71, 73
Display, 25–26, 32, 36
Display applications, 25
Dual damascene structures, 39, 40
Dynamic mechanical analysis, DMA, 178

**E**
Edge dislocation, 68–72
Effective mass, $m_e$, 29, 31, 32, 35
Effective work function, EWF, 3, 17, 18
Elastic coefficients, 7, 9, 20
Elastic modulus, 78–79, 86
Electrical properties, 25–26, 32
Electronic packaging, 93, 98
Electrostatic, 204
Electrostatic potential, 18–19
Elliptical crack growth, 77–78
Embedded atom method, EAM, 56, 68, 80
EPON 862, 215
Epoxy, 149–154, 157, 162, 164, 170, 172–173,
        175, 182–183, 221

Epoxy molding compound, EMC, 151, 196, 233
Epoxy novolac, 233
Epoxy phenol novolac resin, 154
Epoxy/hardener ratio, 152
Epoxy-copper interface, 134
Epoxy-Copper oxide interface
Epoxy-Cu adhesion, 135
Equi-channel angular extrusion, ECAE, 81,
        82, 86
Equilibrium molecular dynamics, EMD, 94
Etch, 39–52
Etch selectivity, 39
Extrinsic grain boundary dislocation, EGBD, 71

**F**
Fence defects, 39
Field to breakdown, FBD, 28, 32
Finite element analysis, FEA, 190, 191, 195
Finite element modelling, FEM, 150
Flat panel display, FPD, 25
Fluorinated silica glass, FSG, 116
Forcefield, 51, 164, 193
Fourier transform infrared spectroscopy, FTIR,
        41, 43, 44, 123
Fractional free volume (FFV), 171

**G**
Generalized gradient approximation, GGA, 7,
        12, 17
Glass transition temperature, $T_g$, 156, 179, 213
Grain boundaries, 55–56, 58, 60, 63–65, 68,
        72, 73
Grain boundary, 81, 86
Grain size, 79–82, 85–87, 89
Grain size hardening, 55, 61
Grain size softening, 55, 62, 64

**H**
Hall-Petch region, 62–64
Hardener, 215–216, 218, 222, 228
Heat flux, 93–94, 96, 102–105
Homogenous theory, 116
Hydrolysis, 41–43, 45, 47–48
Hydrophilicity, 205, 211
Hydrophobicity, 134, 205, 211
Hygroscopic swelling, 155

**I**
Inorganic bottom antireflective coating,
        iBARC, 39, 40, 45, 49, 51

Integrated chip, IC, 77, 78
Integrated circuit package, IC package, 101, 189, 247
Integrated circuits/chip, IC, 247
Interface model, 117
Interfacial adhesion, 134, 138, 143
Interfacial delamination, 189, 191, 195, 199
Interfacial energy, 133–134, 136, 138, 142, 146
Interfacial MD model, 191–192, 197
Interlayer dielectric, ILD, 46, 122, 123
International Technology Roadmap for Semiconductors, ITRS, 77, 78
Intrinsic grain boundary dislocation, IGBD, 70, 71

**J**
Jump in scale, 232

**L**
Langmuir-Blodgett, LB, 204
Large Scale Atomic/Molecular Massively Parallel Simulator, LAMMPS, 14, 80
Leakage current, 25–28, 32
Length scales, 190, 232
Liquid crystal, LC, 26, 32, 36
Local density approximation, LDA, 7, 15
Low cycle fatigue, LCF, 78, 89
Low dielectric materials, 130
Low k, 47
Low k dielectric, 39, 190

**M**
Magnetic properties, 6
Martini Force Field, 234, 233
MD molecular dynamics, 93, 102, 129, 150
Mean free path, 96–97
Mean free path (MFP), 103, 111
Mean squared displacement, 137, 167
Mechanical properties, 115, 122
Mechanical stresses, 93
Mesocite, 232, 238
Mesoscale, 213–249
Mesoscale bead, 219
Misorientation angle, 57–58, 60, 68
Mobility, 27–29, 36
Modulus, 157, 177–178
Moisture, 93–94, 96, 98
Moisture diffusion, 134–137, 141, 144, 146
Moisture diffusion coefficient, D, 141
Moisture simulation, 93, 95, 134, 159

Molecular dynamics, MD, 150
Molecular mechanics MM, 173
Molecular modeling, 25, 39, 101, 117, 163
Molecular modeling strategy, 115, 118
Molecular simulation, 79
Multiscale, 189–191, 195, 198–199
Multiscale approach, 203–204, 211
Multiwalled carbon nanotube, MWCNT, 94

**N**
Nanocrystalline copper, 77, 79, 81, 87–89
N-doped metal oxide semiconductors, NMOS, 18
Non equilibrium molecular dynamics (NEMD), 94, 95, 102, 104
Novolac resin, 154
NPT: constant number of particles, pressure and temperature, 222
Nuclear magnetic resonance, NMR, 123

**O**
Optical properties, 3, 16–17
Organic bottom anti-reflective coating, oBARC, 39
Organosilicate glass, OSG, 116
Organosiloxane, 40

**P**
Particle parameterization, 232
Perdew, Burke, and Ernzerhof (a common form of a GGA), PBE, 7, 11, 31
Periodic boundary conditions (PBC), 12, 28, 55–76, 80, 84, 95, 135–136, 150–151, 164–166, 173–175, 180, 181, 183, 192, 197, 205, 213–249
Periodic cell, 216, 234, 242
Periodic full cell, 237
Periodic vacuum cell, 241
Phenol Novolac EPN 1180
Phonon , 103, 107, 111
Phonon mean free path, 98
Phosphate silicate glass, PSG
Pillar spacing, 209–211
Plasma etch, 39–40, 42, 44, 51
Polarizability, 31, 36
Polycrystalline metals, 55–57, 63, 73
Polymer consistent force field, PCFF, 206
Pore size, 115, 122
Porosity ratio, 116
Positronium annihilation lifetimer spectroscopy, PALS, 162, 171, 172

**P** (*cont.*)
Post ash etchant and cleaner
Pre-crack, 195, 197
Prediction, 3, 8, 22

**Q**
Quantitative structure-property relationships,
  QSPR, 5, 8
Quasicontiuum method, QC, 66, 69, 72

**R**
Radius of a volume-equivalent sphere (res),
  171
Reflectivity, 6, 16–17
Refractive index, 6, 16–18, 22
Relative reactivity, 49
Resistance-capacitance delay, RC delay, 116
Roughened surface, 204
Rule of mixtures, 86, 89

**S**
Self assembled monolayer, SAM, 144, 145, 198
Self assembly, 133, 134, 204
Self assembly monolayer, 133–134
Semi-empirical low cycle fatigue
Sessile drop, 206, 210
Shearing action, 233
Silicon nanocluster, 110–111
Silsesquioxane, 116
Single walled carbon nanotube, SWCNT, 93,
  94, 107, 109
Slip plane, 56, 58, 65, 68, 71
Spin-on-glass, SOG, 26
Stress intensive factor, K, 79
Stress strain curve, 86, 236, 246
Stress-displacement, 195, 197
Structural parameters, 111
Structural properties, 7
Subroutine, 213, 216
Supercell, 14, 18, 233, 235, 247
Surface roughness, 203–204, 209, 211
Symmetrical grain boundaries, 58

**T**
Tapered double cantilever beam, TDCB, 138,
  197
Tensile, 189, 191, 193, 197, 198–199
Tension, 95–98

Thermal conductivity, 5, 7, 14–15, 21, 93–94,
  96–98, 101–105, 108–111
Thermal expansion, 3, 11, 13, 21, 151, 155,
  162, 177, 179
Thermal interface material, TIM, 93
Thermal oxide, ThOx, 51
Thermodynamic properties, 6
Thermo-mechanical properties, 6
Tilt grain boundary, 72
Torsion, 95, 97–98
Transport properties, 21
Tuned (response), 48–49, 51
Twin boundary, 60

**U**
Ultrafine grained metals, UFG, 56
Uncrosslinked, 234, 236, 240, 242, 246
Underfills, 46, 214
Uniaxial fatigue model, 78, 89
Unit cell, 135–140, 145

**V**
Van der Waals, VDW, 174, 233, 234
Via first trench last, VFTL, 39
Vienna Ab initio Simulation Package, VASP,
  7, 11, 17, 19
Virtual internal bond, VIB, 200
Void formation, 246
Voltage holding ratio, VHR, 26, 27, 35, 36
Volume fraction of grain boundaries, 86
Volume of fluid model, VOF, 208
Voronoi tessellation, 80

**W**
Wet etch, 43, 51
Wettability, 203–204, 209, 211
Wetting, 41, 46–47, 49, 51
Work function, 3, 13, 18, 20, 22

**Y**
Yield, 55, 162, 236, 242
Young's equation, 203
Young's modulus, 85, 122, 139

**Z**
Zigzag CNT, 109